Biometric Systems

Edited by Muhammad Sarfraz

Published in London, United Kingdom

IntechOpen

Supporting open minds since 2005

Biometric Systems
http://dx.doi.org/10.5772/intechopen.87715
Edited by Muhammad Sarfraz

Part of IntechOpen Book Series: Biomedical Engineering, Volume 8
Book Series Editor: Robert Koprowski

Contributors

Marat Robert Bogdanov, Anna Filippova, Guzel Shakhmametova, Samma Hussein, Shahrel Azmin Suandi, Xu Liang, David Zhang, Dandan Fan, Zhaoqun Li, Jinyang Yang, Mohd. Abdul Muqeet, Rohit Raja, Hiral Raja, Rajkumar Patra, Akanksha Gupta, K. Ramya Laxmi, Kamal Mehta, Muhammad Sarfraz, Nikolai N. Oskin, Qazi Mateenuddin Hameeduddin

First published in London, United Kingdom, 2021 by IntechOpen
IntechOpen is the global imprint of INTECHOPEN LIMITED, registered in England and Wales, registration number: 11086078, 5 Princes Gate Court, London, SW7 2QJ, United Kingdom
Printed in Croatia

British Library Cataloguing-in-Publication Data
A catalogue record for this book is available from the British Library

Additional hard and PDF copies can be obtained from orders@intechopen.com

Biometric Systems
Edited by Muhammad Sarfraz
p. cm.
Print ISBN 978-1-78984-187-9
Online ISBN 978-1-78984-188-6
eBook (PDF) ISBN 978-1-78984-466-5
ISSN 2631-5343

IntechOpen Book Series

Biomedical Engineering

Volume 8

Muhammad Sarfraz is a Professor and Director of MSIT in the Department of Information Science, Kuwait University, Kuwait. His research interests include computer graphics, computer vision, image processing, machine learning, pattern recognition, soft computing, data science, intelligent systems, information technology and information systems. Prof. Sarfraz has been a keynote/invited speaker at various platforms around the globe. He has advised more than 85 students for their MSc and Ph.D. theses. He has published more than 400 publications as books, journal articles, and conference papers. Prof. Sarfraz is a member of various professional societies. He is the Chair and member of the International Advisory Committees and Organizing Committees of various international conferences. He is also Editor-in-Chief and Editor of various international journals.

Editor of Volume 8:
Muhammad Sarfraz
Department of Information Science
College of Life Sciences, Kuwait University
Sabah AlSalem University City, Shadadiya, Kuwait

Book Series Editor:
Robert Koprowski
University of Silesia, Poland

Scope of the Series

Biomedical engineering is one of the fastest growing interdisciplinary branches of science and industry. The combination of electronics and computer science with biology and medicine has resulted in improved patient diagnosis, reduced rehabilitation time and better quality of life. Nowadays, all medical imaging devices, medical instruments or new laboratory techniques are the result of the cooperation of specialists in various fields. The series of biomedical engineering books covers such areas of knowledge as chemistry, physics, electronics, medicine and biology. This series is intended for doctors, engineers and scientists involved in biomedical engineering or those wanting to start working in this field.

Contents

Preface

A biometric system is a technological system that uses information about a person or other biological organism to identify that person. The biometric industry is rapidly changing and progressing at an astonishing speed. What used to be a futuristic concept has become a reality today. In order to work correctly and effectively, biometric systems depend and rely on data about specific biological traits. In the current age and time, biometric systems are of intensive needs and are widely used in various real-life applications. There are a number of potential applications that have different requirements at different places and times. These, for example, include personal recognition, identification, verification, and others. It may be needed for safety, security, permission, banking, crime prevention, forensics, medical applications, communication, face finding, and others.

The increasing trends, needs, and applications of biometric systems make it more prominent to make developments in this direction to achieve more recent and desired objectives. This leads to the idea of capturing, storing, finding, retrieving, analyzing, and using biometrics in everyday life under the computing environment. Being a computer-based technology, biometric systems carry out automatic processing, manipulation, and interpretation of personal information. It plays a significant and important role in various aspects of real life. It is also highly useful in many areas, disciplines, and fields of science and technology.

This book is specifically dedicated to biometric systems, research, applications, techniques, tools, and algorithms that originate from areas such as image processing, computer vision, pattern recognition, signal processing, artificial intelligence, intelligent systems, soft computing in particular, and in the fields of computer engineering, electrical engineering, and computer science in general. This book explores the latest developments, theories, methods, approaches, algorithms, analyses, and systems for advancements in biometrics and related systems. This publication provides an effective platform for helping and guiding readers, professionals, researchers, academicians, engineers, scientists, and policy makers involved in the area of biometrics. It will disseminate information about backgrounds, methodologies, technologies, and systems in this area together with an in-depth discussion of its latest advances.

The main objective of this book is to provide the international community with an effective platform in the area of people identity verification and authentication from physiological and behavioral aspects. It aims to publish the latest developments and insights for biometric innovations, systems, and applications. The book is also targeted to describe the latest emerging settings and requirements in biometric systems and technologies.

Sarfraz begins the book with an introductory chapter on fingerprint biometrics. He describes that, in the current age and time, biometrics have been deployed successfully in various fields of real life. Numerous methods, techniques, and systems in biometrics serve the sciences, security, military, medical area, and human identification. There are various kinds of biometrics being used, these include fingerprint, face, speaker/voice, infrared thermogram (facial, hand or hand vein), gait, keystroke, odor, ear, hand geometry, retina, iris, palmprint, signature, DNA, knuckle crease, brain/EEG, and heart sound/ECG. In recent years, a large amount of research has been undertaken regarding the evolution of biometric-based information on the fingerprint. Fingerprints are a very effective recording of their superior properties such as reliability and accuracy. This chapter presents the fingerprint recognition concept. It highlights and analyses the work done by different authors related to fingerprint recognition. A detailed comparative study is included in this chapter. It concludes with a discussion of future trends.

This is followed by the chapter titled "Biometric Authentication Based on Electrocardiogram" Bogdanov et al, in this chapter, aim to discuss biometric identification based on electrocardiograms. They state that biometric identification is a multi-stage process, including stages such as signal registration, signal pre-processing, extraction of biometric features, assessment of information content, and selection of the most informative features, as well as classification of biometric features. Each of the stages of biometric identification contributes to the final recognition accuracy. This chapter discusses each of the stages of biometric identification based on ECGs. The results discussed, in this chapter, have been obtained using a computational experiment conducted using programs written in Python. At the same time, the authors have used popular libraries, such as sklearn, scipy, wfdb, biosppy, tensorflow, and others. Most of the input data for the computational experiment has been taken from www.physionet.org. The authors have identified the main factors affecting the accuracy of biometric identification using ECGs. They state that traditional password-based authentication methods have a number of disadvantages related primarily to the human factor. Therefore, biometric methods of identification and authentication are much more reliable, although they have some disadvantages. Some of them, like fingerprints, retina, and voice, was compromised. However, it is not clear what to do if hackers gain access to a biometric database, because a person cannot change fingerprints as easily as a forgotten password. The development of wireless technologies and technologies of the Internet of medical things makes possible the emergence of new biometric identification scenarios.

Muqeet and Hameeduddin, in Chapter 3 of the book, follow with a discussion on "Face Identification using LBP-based Improved Directional Wavelet Transform." They assert that face identification is one of the most active areas of research in computer vision and biometric authentication. It is a well-known fact that various face identification methods have been developed over time. But still, numerous facial appearances including facial expression, pose, and illumination variation, have to be dealt with. Moreover, faces captured in unrestrained situations also impose immense concern in designing effective face identification methods. It is desirable to extract robust local descriptive features to effectively characterize such facial variations both in unrestrained and restrained situations. This chapter discusses such a face identification method that incorporates a popular local descriptor such as local binary patterns (LBP) based on the improved directional wavelet transform (IDW) method to extract facial features. This designed method is applied to complex face databases such as CASIA-WebFace and LFW, which consists of a large number of face images collected under an unrestrained environment with extreme facial variations in expression, pose, and illumination. The chapter describes experiments and includes comparisons with various other methods. These include local descriptive methods as well as local descriptive-based multiresolution analysis (MRA) based methods demonstrating the efficacy of the proposed LBP-based IDW method.

The next chapter, by Liang, et al, is on "Region of Interest Localization Methods for Publicly Available Palmprint Databases". There exist many publicly available palmprint databases that can be found in the current literature. However, not all of them provide the corresponding region of interest (ROI) images. If everyone uses their own extracted ROI images for performance testing, the final accuracy is not strictly comparable. Since ROI localization is the critical stage of palmprint recognition, location precision has a significant impact on the final recognition accuracy. This is very specifically true in unconstrained scenarios. This problem somehow limits the applications of palmprint recognition. However, various published surveys in the current literature focus only on feature extraction and classification methods. Many of the new ROI localization methods have been proposed in recent years. In this chapter, the authors have attempted to group the existing ROI localization methods into different categories. They have analyzed their basic ideas, reproduced some of the codes, made comparisons of their performances, and provided further directions. Hopefully, this chapter would be a useful reference for researchers.

As a representative of biometric technology, palmprint recognition provides a reliable and efficient solution in many authentication scenarios. Palmprint images, even in low resolution, contain rich and discriminative biometric information and have the high antispoof capability, which is suitable for person recognition. Currently, many palmprint acquisition devices have been proposed, but how to design the systems is seldom studied. For example, how to choose the imaging sensor, the lens, and the working distance. Liang et al in Chapter 5, "Image Sharpness based System Design for Touchless Palmprint Recognition", aim to find the relationship between image sharpness and recognition performance. They utilize this information to direct the system design. In this chapter, firstly, the authors introduce the development of recent palmprint acquisition systems, and abstract their basic frameworks to propose the key problems needed to be solved when designing new systems. Secondly, the relationship between the palm distance in the field of view (FOV) and image pixels per inch (PPI) are studied based on the imaging model. They have provided suggestions about how to select the imaging sensor and camera lens. Thirdly, the authors have taken into consideration the image blur and depth of focus (DOF). They have analyzed the recognition performances of the image layers in the Gaussian scale space. Based on this, an image sharpness range is determined for optimal imaging. The experiment results are obtained using different algorithms on various touchless palmprint databases collected using different kinds of devices. These achievements could act as references for new system design.

The second to last chapter of the book is motivated by the Convolutional Neural Network (CNN) on "Transfer Learning of Pre-Trained CNN Models for Fingerprint Liveness Detection." In the current literature, many of the machine learning experts expect that transfer learning will be the next research frontier. Indeed, in the era of deep learning and big data, there are many powerful pre-trained CNN models that have been deployed. Therefore, using the concept of transfer learning, these pre-trained CNN models could be re-trained to tackle a new pattern recognition problem. Samma and Suandi, in this chapter, are aiming to investigate the application of the transferred VGG19-based CNN model to solve the problem of fingerprint liveness recognition. In particular, they have modified the transferred VGG19-based CNN model, re-trained and finely tuned, to recognize real and fake fingerprint images. Moreover, the authors have examined different architectures of the transferred VGG19-based CNN model, including the shallow model, medium model, and deep model. To assess the performances of each architecture, the LivDet2009 database was employed. Reported results indicated that the best recognition rate was achieved from the shallow VGG19-based CNN model with 92% accuracy.

The last chapter is about the "Assessment Methods of Cognitive Ability of Human Brains for Inborn Intelligence Potential using Pattern Recognitions" Raja et al, in this chapter, aim to analyze the scientific study related to fingerprint patterns and the brain lobes. Generally, this method is used to find and develop the inborn potential and personality, especially of children. Every person has inborn potential and personality, which helps to analyze each person's strengths and weaknesses. The work, in this chapter, is based on the analysis. It is mainly for reference purposes for scientific research in the field of Galtian. The statistical study is conducted based on the fingerprint analysis. The study describes that the human brain is divided into two parts: the left hemisphere and the right hemisphere. The fingers of the right hand represent the functions of the left brain and the fingers of the left hand represent the functions of the right brain. The human brain is divided into 10 lobes and each lobe is related with a finger. Each lobe represents different bits of intelligence. The chapter illustrates that the detailed analysis of the fingerprint helps people to find the inborn talents. It provides them with the most appropriate learning habits from a young age and improves learning ability effectively. The vital factor of an individual's intelligence is determined by the neural network connection of brain cells. Cognitive science is a scientific study that helps people to learn about themselves.

Muhammad Sarfraz
Department of Information Science,
College of Life Sciences,
Kuwait University,
Sabah AlSalem University City,
Shadadiya, Kuwait

Chapter 1

Introductory Chapter: On Fingerprint Recognition

Muhammad Sarfraz

1. Introduction

The biometric phrase means life measurement in the Greek language [1]. That is any technique used for measuring biological information for recognition goals called biometric. There are various kinds of biometrics being used, these include Fingerprint, Face, Speaker/Voice, Infrared thermogram (facial, hand or hand vein), Gait, Keystroke, Odor, Ear, Hand geometry, Retina, Iris, Palmprint, Signature, DNA, Knuckle crease, Brain/EEG, Heart sound/ECG. Defining humans using biometric can even be behavioral or physiological biometrics. The difference between them is that behavioral biometric can be affected with the progress of the time such as signature, gait, speech, and keystroke but the physiological biometric are constant during human life. Fingerprint, face, iris, and palmprints are examples of physiological biometric [2]. A Biometric system is reliable because it cannot be stolen, borrow, bought, or forgotten like a password or ID [3].

The fingerprint is a physical biometric aspect. It is used to identify a person's identity due to its uniqueness where no two persons can share the same fingerprint. Besides, a fingerprint is unchangeable with time and can be easily recognized during the whole life of the individual. The fingerprint is an impression or model of ribs and valleys at the top of a person's fingers. **Figure 1** shows a fingerprint pattern. Fingerprint recognition is the automatic prosses of comparing saved fingerprint pattern with the input fingerprint to determine human characters. Although fingerprint recognition was deployed from decade it became one of the most common biometric nowadays. The fingerprint identification system is a cheap but solid mechanism at the same time. Moreover, it's a simple way to identify humans speedily and accurately [4]. Many applications applied fingerprint recognition such as the military, judiciary, health, teaching, civic serving, mobiles and laptop log-in, and many more. Modern techniques and approaches are used recently as a substituted of old ink to capture the fingerprint. These technologies differ

Figure 1.
A fingerprint model.

in terms of accuracy, effectiveness, speed, advantages, and challenges [5]. This chapter discusses, compares and analyses several authors work [1–27] regarding the fingerprint recognition.

The remaining of the chapter is organized as follows. Section 2 is an overview of the literature survey with comparative research. Section 3 deals with a detailed analytical study of the literature review. At last, future directions, recommendations, and conclusion are presented in Section 4.

2. Literature survey

Fingerprint recognition is the procedure of comparing known and unknown fingerprints to prove that the it is from the same person or not [8]. Today, many approaches, techniques, and systems are used to match fingerprints and solve related problems. This section is focused on analyzing and categorizing different author's work in the fingerprint recognition area. **Table 1** provides a summary of various papers in the current literature. First column determines the Reference of the papers by author names and year of publication. Second column gives the summary of the work in the corresponding paper, and the third column describes the implemented approaches used to solve fingerprint recognition issues. The author names and the year of publication will be used as an identifier for the rest of the tables in the chapter showing other details of the referred literature.

Reference	Brief summary	Approaches adopted
[8]	Explains different biometrics structures that are used for certification and recognition purpose with submitting their advantages and disadvantages.	• Knowledge-based approach • Token based approach • Biometric based approach
[4]	A general explanation of various types of fingerprint recognition systems and patterns depending on the minute-based technique. Focused on Pattern recognition, wavelet, and wave atom mechanisms. Complications related to the wave atom method are studied.	• Histogram Equalization • Band pass Filtering • Gabor Filtering • Binarization and Thinning • 2D Fourier Transform • Wavelet based Transformation • Wave atom Transform and MCS optimization algorithm
[20]	Explains the differences between various fingerprint matching techniques particularly local minutiae-based matching algorithms. It provides an experiment about fingerprint identification and authentication using the minutiae-based matching method with analyzing the outcomes.	• topology of local structure • type of consolidation • usage of additional features • minutiae peculiarities • parameter learning.
[23]	Discuses fingerprint authentication using minutiae extraction technique and covering all related systems and processes.	• Load image • Histogram Equalization • Fast Fourier Transformation • Binarization • Region of Interest • Thinning • Minutiae Extraction • False Minutiae Removal

Reference	Brief summary	Approaches adopted
[21]	Beneficent of minutiae-based fingerprint verification system by suggesting a route for the feature extraction step which depends on reexamining the gray-scale profile can increase the matching performance by 4%. Also, the proposed feature refinement step that allocates class labels for every 31qmintiae will improve the performance by 3%. Both steps will develop the whole fingerprint verification system be 8%.	Sequential approach
[18]	Execution and assessment of Biometric Image Software (NBIS) for fingerprint recognition developed by the National Institute of Standards and Technology (NIST). the NBIS is implemented in the MATLAB environment.	• Pre-processing • Minutiae Extraction • Post processing
[17]	Design minutia extractor by using different techniques. Some improvements in the thinning, false removal approach, and image segmentation is implemented in the work.	• Segmentation using Morphological operations • Thinning • False minutiae removal methods • Minutia marking • Minutia unification by decompose-ng a branch into three termination • Matching in the unified x-y coordinate system
[24]	Combining minutia and correlation-based approaches to evolve an automatic fingerprint recognition system. By using this hybrid, the performance of the minutiae algorithm is grown.	• Minutiae Extraction • Post-processing • Minutiae Matching • Filtering • Feature Vector
[2]	Present Fingerprint Recognition using the Minutia Score Matching method (FRMSM). It implements Block Filter for fingerprint thinning. Also, it compares with available algorithms.	• Thinning • Image binarizing • Noise removal
[1]	A summary of several biometrics techniques as well as explaining the unimodal and multimodal with their pros and cons.	• Sensor module • Matching module • Decision-making module • Feature extraction module
[3]	Explaining some biometrics and dividing them to currently in use biometrics, limited used biometrics, and understudy biometrics.	Fusion scheme
[15]	An alignment-based minutia-matching algorithm has been developed to increase the speed and accuracy by ability determining the matches between input minutiae and Stord one without the need for detailed study. Michigan State University and the National Institute of Standards and Technology NIST 9 fingerprint databases have been used. The result shows that the full verification process takes 1.4 seconds a Sun ULTRA 1 workstation.	Alignment-based minutiae-matching algorithm

Reference	Brief summary	Approaches adopted
[22]	Applying fingerprint identification by employing a gray level watershed process to find out the ridges present on a specific fingerprint image. The result display that this system is accurate and fast when matching 7 images in the database.	• Image acquisition • Preprocessing • Minutiae detection • Minutiae reduction • Fingerprint matching
[26]	Discussing fingerprint recognition biometric in detail and explaining deferent types of algorithms like negative Laplace filter and the non-stationary analysis, and a flexible algorithm with calculating the matching test results.	• Image acquisition • Preprocess-ng • Segmental-on • Minutia detection • Biometric matching
[10]	Developing a novel algorithm for fingerprint matching based on local structures to elicit neighboring minutiae features effectively. The presented algorithm is tested on FVC2002 and the results show the reliability of the system.	Novel topology-based representation technique
[9]	Mixing the density map matching with minutiae-based matching where the density data can be used in the matching process to reduce extra storing cost. The outcomes approved that combining both approaches will improve performance.	• Region estimation • Orientation filed estimation • Fingerprint enhancement • Coarse density map extraction • Weighted polynomial approximation
[12]	An adequate wat to press the template size with a reduction ratio of 94% by applying tow reduction algorithms the Column Principal Component Analysis and the Line Discrete Fourier Transform feature reductions. Also, a fast minutiae-based matching algorithm can be accomplished throw spectral minutiae fingerprint recognition system which shows matching speed with 125000 comparisons per second on a PC with Intel Pentium D processor 2.80 GHz and 1 GB of RAM.	• Column Principal Component Analysis (Column-PCA) • Line Discrete Fourier Transform (Line- DFT)
[25]	Novel core point detection method that uses the detection algorithm to examine the core point and determine local frame for minutiae close to it. Then tow fingerprint corresponding points will be earned and used to match the global class then make the final diction.	Core-based structure matching algorithm
[7]	New topology-based algorithms to match fingerprint and address the local matching, tolerance to deformation, and global matching. The experiment outcomes approve that time and performance is improved using the algorithm.	Topology-matching algorithm
[16]	Provide a hybrid matching algorithm that matches fingerprints using minutiae inputs and texture inputs together. The matching performance improved when testing 2560 images by collecting both texture-based and minutiae-based matching scores.	hybrid matching approach (minutiae-based representation with a texture-based representation)
[19]	Suggesting ridge feature-based approach for fingerprint recognition that provides good results for low-quality fingerprint images. Matching fingerprint images based on ridgeline features extracted by using contextual filtering and two pass thinning. Histogram approach is used to match the fingerprint. The experiments show how the performance developed using this approach.	• Contextual filter • Single pass thinning algorithm • Image preprocess • Gabor filtering

Reference	Brief summary	Approaches adopted
[13]	Novel enhancement algorithm that split the input fingerprint image to set of filtered images which will help in producing orientation field and quality mask. The evaluation process of the algorithm is done on an online fingerprint verification system using the MSU fingerprint database that consists of 600 fingerprint images and the test demonstrates that the enhancement algorithm improves the performance of the online fingerprint verification system.	• Gabor filters • Ridge extraction algorithm • Voting algorithm • Orientation estimation algorithm
[14]	Submit a fingerprint recognition algorithm depending on phase-based image matching. Which uses the phase components in 2D (two-dimensional) discrete Fourier transforms of fingerprint images to reach strong fingerprint recognition with a low-quality fingerprint. The test used a group of fingerprint images captured from fingertips with a bad case. The results show an effective recognition performance using this approach.	2D (two-dimensional) Fourier transforms
[6]	The correlation-based fingerprint verification system uses the richer gray-scale information of the fingerprints. In the beginning, the system chooses appropriate templates in the primary fingerprint, employs template matching to locate them in the secondary print, and match the template positions of both fingerprints. The test describes the performance of correlation-based fingerprint against other systems.	• Classification of template positions • Elementary decisions • Combining elementary decisions
[5]	A brief summary of fingerprint matching techniques, systems, and performance evaluation.	• Image capturing module • Feature extraction module • Pattern matching module
[11]	It provides important aspects of fingerprint recognition. As biometric pattern, it highlights a detailed analysis on the fingerprint conceptualization. It uses various tools to find the match percentage in the verification process.	• Negative Laplace filter • Non-stationary analysis of the short time Fourier transform • An algorithm to find the match percentage in the verification process.
[27]	This presents a fast fingerprint enhancement algorithm, which can adaptively improve the clarity of ridge and valley structures of input fingerprint images based on the estimated local ridge orientation and frequency.	• Goodness index of the extracted minutiae • Accuracy of an online fingerprint verification system.

Table 1.
Overview of the literature.

Table 2 shows the accuracy and performance in percentage. It also mentions the identification and verification measures. Identification and verification are matching techniques for fingerprint recognition. In the verification, the person enrolls his fingerprint to the system and the templet stored it in the database. Every time the person accesses the system, he has entered his fingerprint to verify himself. It's a one to one relationship where the input fingerprint is compared with the stored one. On the other hand, identification is one to many relationships because the human fingerprint is matched with the fingerprints database to determine who is that person [8].

Reference	Accuracy (Performance)	Performance measures used for verification	Performance measures used for identification
[8]	_	_	_
[4]	_	FAR, FRR, FMR, FNMR, ERR	Accuracy
[20]	_	FMR, FNMR, EER, ROC, FMR100, FMR1000, Zero FMR	True positive rate (TPR), R100, ZeroR, Cumulative Match Curve (CMC), Accuracy, computational time, rank k
[23]	_	FMR	_
[21]	95% (LVQ-based classifier on training data) 87% (LVQ-based classifier on test data)	FAR, GAR	Classification accuracy,
[18]	_	FNMR, FMR	Reliability and quality
[17]	_	FRR, FAR	Quality and accuracy
[24]	_	FAR	_
[2]	_	FMR, FNMR	_
[1]	_	FMR, FNMR, FTC, FTE	accuracy, speed, resource requirements, acceptability, and circumvention.
[3]	_	_	_
[15]	_	FAR, FRR	Accuracy, speed
	More than 45%	_	Accuracy and testing time.
[26]	_	false acceptance (FA), false rejection (FR), recognition rate (RR)	Accuracy
[10]	_	EER	_
[9]	_	FAR, FRR	Matching time and computation cost
[12]	_	FAR, EER, GAR	Recognition accuracy, matching speed and robustness to poor image quality
[25]	_	FAR, FRR	Matching time
[7]	_	FRR, FAR	Matching accuracy Matching time Computing time
[16]	_	GRA, FAR	Computing time
[19]	98%	EER, FAR, FRR	Matching accuracy
[13]	_	_	Reject Rate Recognition Rate
[14]	_	EER, ZeroFMR, FNMR, FMR	Accuracy
[6]	_	FRR, FAR,FNMR	Testing time
[5]	_	EER, FAR, FRR	_
[11]	_	_	_
[27]	_	_	_

Table 2.
Accuracy and performance.

While the performance measures used for identification depend on the accuracy, recognition rate, rank K, etc., the performance measures for verification are False Match Rate (FMR), False Non-Match Rate (FNMR), False Accept Rate (FAR), False Rejection Rate (FRR), and Equal Error Rate (EER). The researchers in [4] describe the meaning of the authentication parameters. FAR happens when the system recognizes person erroneous. But when the system rejects entry to approve person that means the FRR is happening. FMR is the amount of fraud assessments with threshold value 'T' divided by the total quantity of fraud similarities. FNMR is the quantity with unaffected comparisons with threshold value 'T' divided by the total quantity of open comparisons. Last one is EER, it describes the error rate of the system.

The experimental parts of the author's [1–27] are shown in **Table 3**. It explains the type of applications and kind of Databases used. Then it shows the number of

Reference	Application	Database	No. of identities	Total No. of images	Resolution	Image format
[8]	_	_	_	_	_	_
[4]	Fingerprint	_	_	_	_	_
[20]	Fingerprint	FVC	308	1228	_	_
[23]	Fingerprint	_	_	_	_	_
[21]	Fingerprint	IBM HURSLEY database	269	900	500dpi	_
[18]	Fingerprint	FVC 2000	60	480	_	_
[17]	Fingerprint	_	_	2	_	_
[24]	Fingerprint	Biometric System Lab (University of Bologna - ITALY) Ink and scanner	21 7	168 56	256 × 256 × 256dpi 240× 240× 256dpi	_
[2]	Fingerprint	_	_	_	_	_
[1]	• Fingerprint • Face • Voice • Infrared thermogram (facial, hand or hand vein) • Gait • Keystroke • Odor • Ear • Hand geometry • Retina • Iris • Palmprint • Signature • DNA	FVC2002 FRVT2002 NIST2000	_	_	_	_

Reference	Application	Database	No. of identities	Total No. of images	Resolution	Image format
[3]	• Fingerprint • Face • Iris • Hand geometry • Palmprint • Speaker/voice • Signature • Ear shape • Knuckle crease • Brain/EEG • Heart sound/ ECG	_	_	_	_	_
[15]	Fingerprint	MSU fingerprint data base NIST 9 (card 1) NIST 9 (card 2)	70 1350 1350	700 900 900	640 X 480 832 X 768 832 X 768	_
[22]	Fingerprint	Scanner or inked impression	_	7	250 X 250 pixels	TIF and BMP
[26]	Fingerprint	commercial databases	40	_	300 x 300 512 DPI	_
[10]	Fingerprint	FVC2002	400	3200	_	_
[9]	Fingerprint	THU database FCV2002	827 100	6616 800	320X512	_
[12]	Fingerprint	MCYT FVC2002-DB2	145 40	1740 400	_	_
[25]	Fingerprint	Live fingerprint database	_	8000	300*300	_
[7]	Fingerprint	fingerprint database at University of Bologna, Italy	21	1680	256 × 256	_
[16]	Fingerprint	_	160	2560	_	_
[19]	Fingerprint	NRC FVC2000 database	_	300	200 × 200	_
[13]	Fingerprint	MSU	67	670	640*480	_
[14]	Fingerprint	_	30	330	256 × 384	_
[6]	Fingerprint	FVC2000	110	880	_	_
[5]	Fingerprint	_	_	_	_	_
[11]	Fingerprint	_	_	_	_	_
[27]	Fingerprint	_	_	_	_	_

Table 3.
Overview of the used data.

Reference	Methods used	Reason of application	Advantages	Disadvantages
[8]	–	–	–	–
[4]	• Minutiae based approach • Pattern Recognition Approach • Wavelet based Approaches	• To compare the fingerprint patterns. • The use of patterns for authentication purpose • Used on fingerprint pattern to carry out the verification.	• Great accuracy rate.	• Image with noise or encrypted cannot be used, slow approach and fails to determine real humans. • Not required finger printing or post processing, work in the least three levels of texture split to make the system excellent and its fast process.
[20]	• Minutiae-based local matching • Correlation-based matching techniques • Indexing algorithms	• Comparing tow fingerprints to gain a result of matching or nonmatching. • Calculate the similarities between tow fingerprint images by the correlation within corresponding. • Used when it's important to enter fast to the fingerprint templates for recognition.	• Simple and distortion tolerance. • Simplicity	• Expensive computation, slow and depend on the skin situation.
[23]	Minutiae based matching	Minutia extracted from fingerprint and saved in the database then the matching happened between the stored and input fingerprint.	Widely used and familiar.	Affected with the wet or dry skin.
[21]	minutiae-based fingerprint verification system	• Resolve the gray scale profile in the neighbor-hood of potential minutiae. • Understand the gray level image properties.	–	–
[18]	Biometric Image Software (NBIS)	Used for fingerprint recognition in MATLAB environment	–	Time consuming, bad performance for images.
[17]	Minutiae Extraction Technique	Used to reduce distortion for fingerprint matching.	Reduce execution time.	–
[24]	hybrid Automatic Fingerprint Recognition System (Hybrid APRS)	Hybrid between minutiae and correlation-based techniques to represent and match fingerprint.	• Improve each technique individually. • Improve minutia algorithm. • improve the ridge algorithm.	–

Reference	Methods used	Reason of application	Advantages	Disadvantages
[2]	Minutia Score Matching method (FRMSM)	Matching the input fingerprint with the stores fingerprint database.	–	–
[1]	• Unimodal biometric systems • multimodal biometric system	• Using one single biometric feature. • Using various applications to benefit from different types of biometrics advantages	• Reliability due to use the combination of deferent biometric strength.	• Scanned data became noisy. • Varity in the level of difficulty in the data gained from humans. • There may be a lot of similarity in the features sets of the used biometric. • Some individuals may not have the chosen biometric crater. • Biometric sign can expose to forgery.
[3]	• unimodal biometric systems • multimodal biometric system	• Recognition using only one biometric crater. • Recognize person using more than one biometric property.	• Late to progress in the performance.	• Not universal • Can be faceable • Contain many noises • variations within the class. • similarities between the classes
[15]	Automatic identity-authentication system	Use the fingerprint to identify person identity.	Its intended mainly for forensic applications account for approximately $100 million from the world market.	–
[22]	Edge Detection	To find the ridges existed in the fingerprint image	–	–
[26]	Open algorithm system	–	–	–
[10]	Minutiae matching approach	For creating minutiae descriptor	–	–
[9]	Density map matching and minutiae-based matching	Identify the fingerprint ridges denseness and sparseness	• Low storage cost. • major factor for fingerprint representation. • No redundancy between both systems.	–

DOI: http://dx.doi.org/10.5772/intechopen.95630

Reference	Methods used	Reason of application	Advantages	Disadvantages
[12]	Spectral minutiae fingerprint recognition system	Used to represent a minutia set as a fixed-length feature vectcr	• High speed operations. • Low matching time. • Suitable for large scale fingerprint identification system.	–
[25]	• Structure-based matching algorithms • Core-based matching algorithms	– –	• More effective algorithm	• Not suitable for online applications and require long time. • Highly depends on core point detection precision
[7]	Minutiae- based matching	For matching the fingerprints to find the similarities between them.	Good matching capability	• The missing minutiae should be considered. • High cost process. • Hard to nonlinear deformations of fingerprints
[16]	Minutiae- based matching algorithms	–	–	Not enough corresponding points in the input images.
[19]	Ridge feature-based approach	Uses the ridges to match two fingers.	• Need little processing. • Increase matching accuracy. • Powerful with low quality fingerprint images.	–
[13]	Online fingerprint verification system.	–	–	• Slow. • Fail to devolve the clarity of ridges structure for good quality fingerprint templet.
[14]	Phase-based image matching	–	Good results when using bad condition fingertips.	–

Reference	Methods used	Reason of application	Advantages	Disadvantages
[6]	Correlation-based fingerprint verification system	To match tow fingerprint depending on gray level fingerprint images.	Work well with bad quality fingerprint image.	–
[5]	• Minutiae-based matching • Pattern matching	–	–	–
[11]	Fingerprint verification system	–	–	–
[27]	• Minutiae-based matching • Pattern matching	Uses the ridges and valley structures of input fingerprint images.	Improves the goodness index and the verification accuracy	–

Table 4.
Applications used with the advantages and disadvantages.

fingertips used to capture the fingerprints databases, the number of images resulted from the fingers, their resolutions and formats. Finally, **Table 4** describes the implemented application type and the reason for using it by mentioning the advantages and disadvantages of the proposed methods.

3. Data analysis

This section analyses the fingerprint recognition data resulting from the literature [1–27] survey in Section 2. In general, fingerprint recognition processes can be done using multiple procedures. First, decompose raw human fingerprint sample to create digit presentation of the same sample. On the next step, preprocessing is done for the raw input image by filtering and improving fingerprint image to produce suitable output image for feature extraction which extracts the unique features of the fingerprint from the digital representation sample. These extracted features are saved in the fingerprint database as features. Final step is to match the input fingerprint with fingerprint template stored in the database to find the similarities. The outcome of these procedures is deciding if the person is identified or not [8]. **Figure 2** describes the sequence of biometric or fingerprint system. The fingerprint procedures involve many different approaches and algorithms that are used to enhance and improve the low quality of fingerprint images. If the fingerprint image is on good quality, then there are no issues and will appear while matching [4]. **Table 1** presents the approaches that are used by different authors. **Figure 3** presents the most used approaches. Different matching approaches are used in 15 papers which can be considered as the commonly used approaches. Then minutiae extraction techniques are used in around 10 papers. Post processing and histogram equalization are used in 2 papers. There are some other approaches used only once in some of the papers.

When the matching process is completed. Correctness of a fingerprint identification system is calculated by applying some parameters. It is used to measure the performance of identification and verification. The performance measures used for identification depend mostly on the accuracy, testing time and image quality. **Figure 4** confirms that 38% of the work used the accuracy as the main identification measure and applied it alone or in addition to other measures. On the other

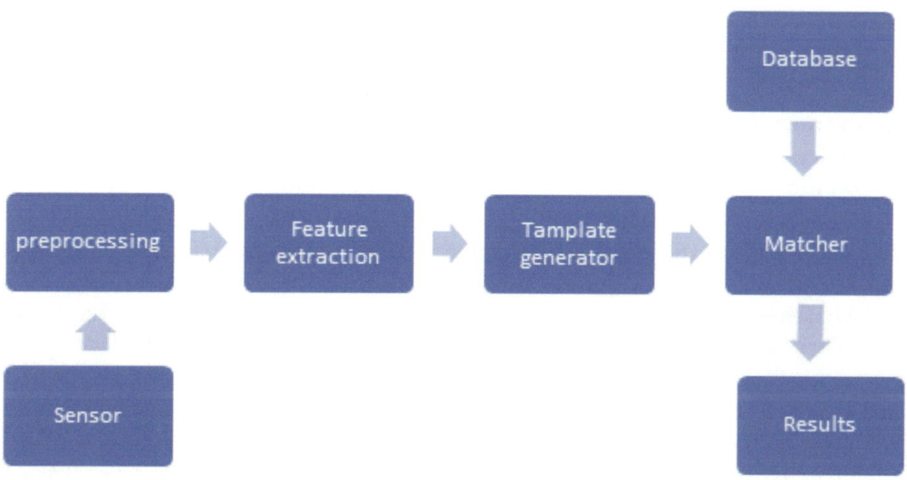

Figure 2.
Biometric or fingerprint system.

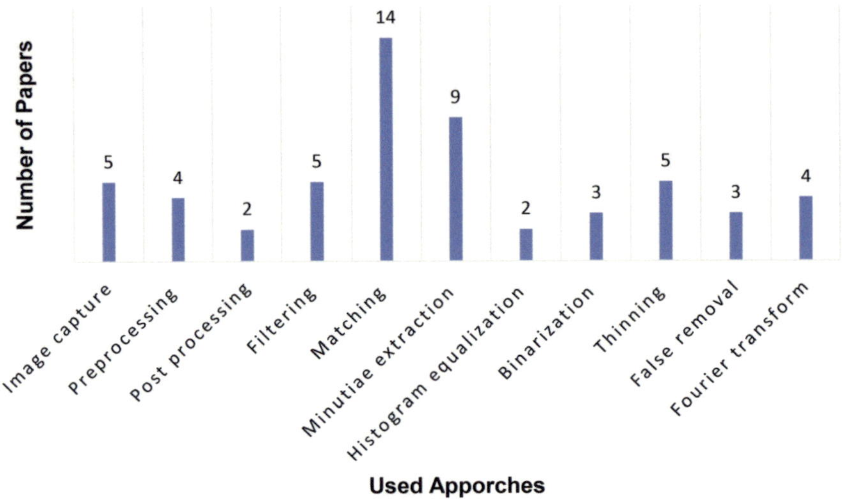

Figure 3.
The most used fingerprint approaches in various papers.

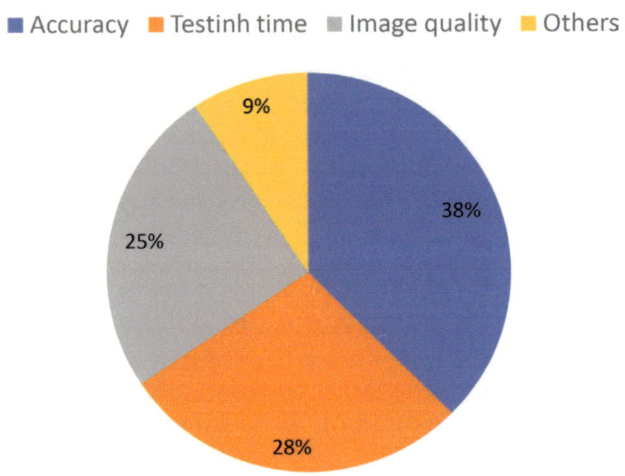

Figure 4.
The identification measures used in the work.

hand, the most applied performance measures for verification are False Match Rate (FMR), False Non-Match Rate (FNMR), False Accept Rate (FAR) and False Rejection Rate (FRR). As shown in **Figure 5**, approximately 36% of the papers rely on (FAR) as a verification measure.

In the fingerprint recognition area, conducting test and experiments is important to approve and evaluate the quality and accuracy of the proposed work. Many different data bases have been used to test the performance of the proposed matching algorithms. These databases vary in their sizes, average number of templets and input fingerprints. **Figure 6** describes the databases types used in the study. As noticed from **Table 3**, FVC2000 and FVC2002 databases are used in some papers but most papers used their own databases. For example, authors in [24] used Biometric System Lab (University of Bologna – Italy). The used databases contain a several number of fingerprints that are used to produce fingerprint images. These images are used in matching step. **Figure 7** shows the discerption of the used databases characteristics by presenting the number of identities and the number of images.

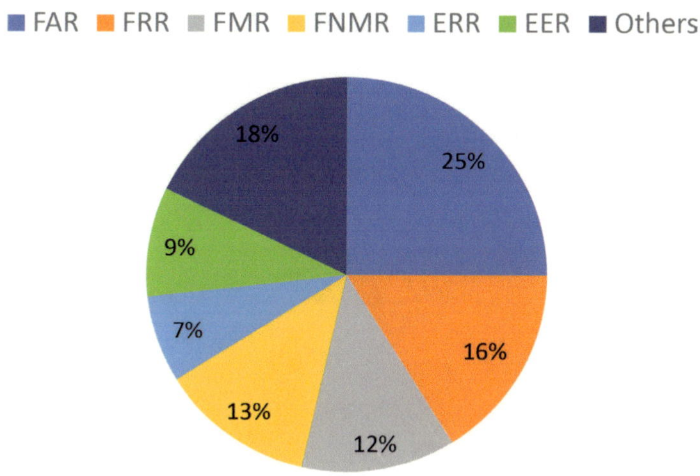

Figure 5.
The verification measures used in the work.

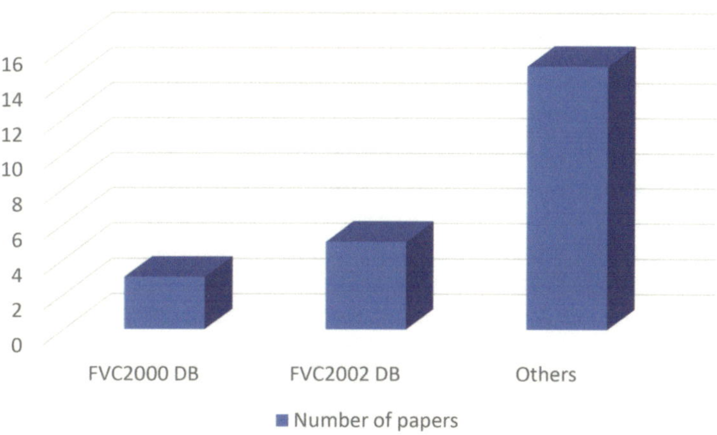

Figure 6.
The used databases in the papers.

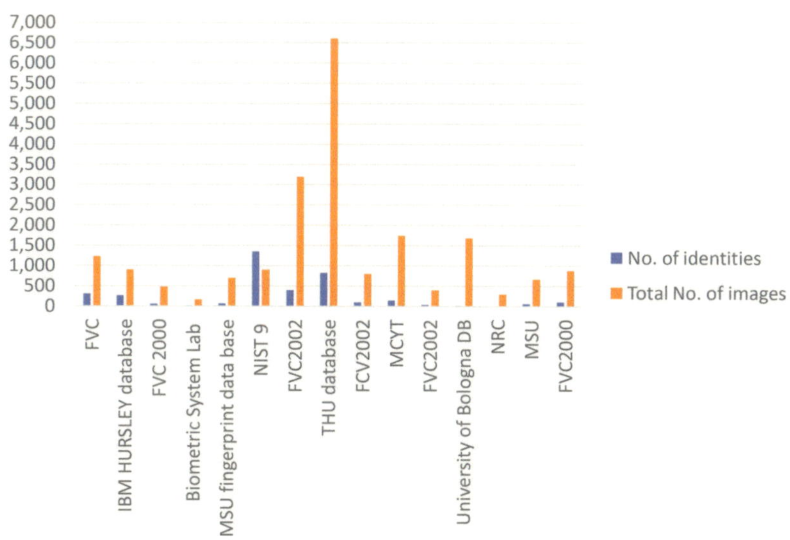

Figure 7.
Used database characteristics.

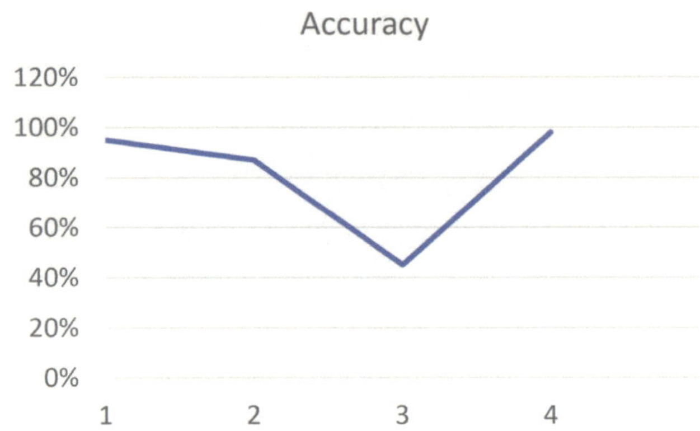

Figure 8.
Accuracy and performance from the used papers.

At last, the evaluation of the performance or accuracy of the fingerprint verification system are appearing in 4 papers as presented in **Figure 8**. The figure shows the highest accuracy with 95% and the lowest accuracy with 45%.

4. Conclusion

Biometrics means the automatic identification of a person based on his behavioral and/or physiological unique characteristics. Fingerprint biometrics is an efficient, safe, cost-effective, easy to use the technique for identity verification. This study provides detailed information related to fingerprint recognition techniques. Several author's works, related to fingerprint recognition technology, are discussed, compared and analyzed. A detailed analysis of various studies is made. As a future work, there is a scope to improve the problems related to fingerprint recognition, specially, the issues related to the capturing row fingerprint by the sensors. One of the innovations is the touchless fingerprint sensor, which will be sufficient for current (COVID-19) situations. It will decree the need to touch the devices. This technique is needed to show its reliability and efficacy as an alternative to regular sensors. Relying on a fingerprint recognition in a different government domains is also recommended. Implementing fingerprint recognition technology is not only useful for Government, but other organizations and communities can also think and may benefit by applying fingerprint recognition techniques to identify. For example, in the health sector, it is quite important to use fingerprint recognition to identify the person injured in an accident.

Author details

Muhammad Sarfraz
Department of Information Science, College of Life Sciences, Kuwait University, Sabah AlSalem University City, Shadadiya, Kuwait

*Address all correspondence to: prof.m.sarfraz@gmail.com; hawra.alhussain@grad.ku.edu.kw

References

[1] Delac, K., & Grgic, M. (2004). A Survey of Biometric Recognition Methods. *Proceedings. Elmar-2004. 46th International Symposium on Electronics in Marine*, Zadar, Croatia, 2004, pp. 184-193.

[2] Ravi, J., K, R. B., & Venugopal, R. K. (2009). Fingerprint Recognition using Minutia Score Matching. International Journal of Engineering Science and Technology, Vol. 1(2), 35-42.

[3] Mir A.H, Rubab, S and Jhat, Z. A. Biometrics Verification: a Literature Survey. Journal of Computing and ICT Research, Vol. 5, Issue 2, pp 67-80. http://www.ijcir.org/volume5-number2/article7.pdf

[4] Borra, S. R., Reddy, G. J., & Reddy, E. S. (2016). A broad survey on fingerprint recognition systems. 2016 International Conference on Wireless Communications, Signal Processing and Networking (WiSPNET), 1428-1434. https://doi.org/10.1109/WiSPNET.2016.7566372

[5] Subban, R., & Mankame, D. P. (2013). A Study of Biometric Approach Using Fingerprint Recognition. Lecture Notes on Software Engineering, 209-213. https://doi.org/10.7763/LNSE.2013.V1.47

[6] Bazen, A., Verwaaijen, G.T., Gerez, S., Veelenturf, L.P., & Zwaag, B.J. (2000). A correlation-based fingerprint verification system. Proceedings of the ProRISC/IEEE workshop, November 30–December 1, 2000, 205-213, ISBN: 90-73461-24-3.

[7] Chengfeng Wang, Gavrilova, M., Yuan Luo, & Rokne, J. (2006). An efficient algorithm for fingerprint matching. 18th International Conference on Pattern Recognition (ICPR'06), 1034-1037. https://doi.org/10.1109/ICPR.2006.236

[8] Deokar, S., & Talele, S. (2014). Literature Survey of Biometric Recognition Systems. International Journal of Technology and Science, 1(2).

[9] Dingrui Wan, & Jie Zhou. (2006). Fingerprint recognition using model-based density map. IEEE Transactions on Image Processing, 15(6), 1690-1696. https://doi.org/10.1109/TIP.2006.873442

[10] Gao, Z., You, X., Zhou, L., & Zeng, W. (2011). A novel matching technique for fingerprint recognition by graphical structures. 2011 International Conference on Wavelet Analysis and Pattern Recognition, 77-82. https://doi.org/10.1109/ICWAPR.2011.6014495

[11] Gonzalez, F. C. J., Villegas, O. O. V., Sanchez, V. G. C., & Dominguez, H. d. J. O. (2010). Fingerprint Recognition Using Open Algorithms in Frequency and Spatial Domain. 2010 IEEE Electronics, Robotics and Automotive Mechanics Conference, 469-474. https://doi.org/10.1109/CERMA.2010.117

[12] Haiyun Xu, Veldhuis, R. N. J., Kevenaar, T. A. M., & Akkermans, T. A. H. M. (2009). A Fast Minutiae-Based Fingerprint Recognition System. IEEE Systems Journal, 3(4), 418-427. https://doi.org/10.1109/JSYST.2009.2034945

[13] Hong L., Jain A. (2004) Fingerprint Enhancement. In: Ratha N., Bolle R. (eds) Automatic Fingerprint Recognition Systems. Springer, New York, NY. https://doi.org/10.1007/0-387-21685-5_7

[14] Ito, K., Morita, A., Aoki, T., Higuchi, T., Nakajima, H., & Kobayashi, K. (2005). A fingerprint recognition algorithm using phase-based image matching for low-quality fingerprints. IEEE International Conference on Image Processing 2005, II–33. https://doi.org/10.1109/ICIP.2005.1529984

[15] Jain, A. K., Lin Hong, Pankanti, S., & Bolle, R. (1997). An

identity-authentication system using fingerprints. Proceedings of the IEEE, 85(9), 1365-1388. https://doi.org/10.1109/5.628674

[16] Jain, A., Ross, A., & Prabhakar, S. (2001). Fingerprint matching using minutiae and texture features. Proceedings 2001 International Conference on Image Processing (Cat. No.01CH37205), 2, 282-285. https://doi.org/10.1109/ICIP.2001.958106

[17] Kaur, M., Singh, M., Girdhar, A., & Sandhu, P. S. (2008). Fingerprint Verification System Using Minutiae Extraction Technique. 2(10), 6.

[18] Maddala, Sainath, et al. "Implementation and Evaluation of NIST Biometric Image Software for Fingerprint Recognitionc." ISSNIP Biosignals and Biorobotics Conference: Biosignals and Robotics for Better and Safer Living, BRC, 2011.

[19] Mar Win, Z., & Myint Sein, M. (2011). An Efficient Fingerprint Matching System for Low Quality Images. International Journal of Computer Applications, 26(4), 5-12. https://doi.org/10.5120/3094-4246

[20] Peralta, D., Galar, M., Triguero, I., Paternain, D., García, S., Barrenechea, E., Benítez, J. M., Bustince, H., & Herrera, F. (2015). A survey on fingerprint minutiae-based local matching for verification and identification: Taxonomy and experimental evaluation. Information Sciences, 315, 67-87. https://doi.org/10.1016/j.ins.2015.04.013

[21] Prabhakar, S., Jain, A. K., Jianguo Wang, Pankanti, S., & Bolle, R. (2000). Minutia verification and classification for fingerprint matching. *Proceedings of the 15th International Conference on Pattern Recognition. ICPR-2000*, Barcelona, Spain, pp. 25-29, Vol.1, doi: 10.1109/ICPR.2000.905269.

[22] G.S. Rao, C. NagaRaju, L.S.S. Reddy, & E.V. Prasad. (2008). A Novel Fingerprints Identification System Based on the Edge Detection. IJCSNS International Journal of Computer Science and Network Security, Vol. 8(12), 394-397.

[23] Sharma, M. (2014). Fingerprint Biometric System: A Survey. International Journal of Computer Science & Engineering Technology, Vol. 5(7), 743-747.

[24] Youssif, A. A. A., Chowdhury, M. U., Ray, S., & Nafaa, H. Y. (2007). Fingerprint Recognition System Using Hybrid Matching Techniques. 6th IEEE/ACIS International Conference on Computer and Information Science (ICIS 2007), 234-240.

[25] Zhang, W., Wang, S., & Wang, Y. (n.d.). Core-Based Structure Matching Algorithm. 10.

[26] F. C. J. González, O. O. V. Villegas, V. G. C. Sánchez and H. d. J. O. de Jesús Ochoa Dominguez, "Fingerprint Recognition Using Open Algorithms in Frequency and Spatial Domain," *2010 IEEE Electronics, Robotics and Automotive Mechanics Conference*, Morelos, 2010, pp. 469-474, doi: 10.1109/CERMA.2010.117.

[27] Lin Hong, Yifei Wan and A. Jain, (1998). Fingerprint image enhancement: algorithm and performance evaluation, *IEEE Transactions on Pattern Analysis and Machine Intelligence*, 20(8), pp. 777-789, doi: 10.1109/34.709565.

Chapter 2

Biometric Authentication Based on Electrocardiogram

M.R. Bogdanov, A.S. Filippova, G.R. Shakhmametova and Nikolai N. Oskin

Abstract

The life of modern society is impossible without trust. To ensure trust in the digital world, various encryption algorithms and password policies are used. Passwords are used in a variety of applications from banking applications to email. The advantages of passwords include ease of use and widespread distribution. Forgotten password can be restored or changed. Password weaknesses are largely related to the human factor. Many users use passwords such as "1234" or "qwerty," and they are also willing to share passwords with friends and colleagues. Vulnerabilities are also associated with software and hardware manufacturers. Many Wi-Fi routers preset very simple passwords, which many users leave unchanged. There are questions for manufacturers of mobile applications. Due to the imperfection of their software, personal data of users often leak. Due to the prevalence of social networks, new authentication methods have appeared. On many websites, you can use accounts from Facebook or Gmail.com for authentication. If hackers manage to break into large IT vendors, then millions of accounts will be leaked. Many common password problems can be overcome with biometric identification. In particular, biometric data are very difficult to fake; they usually do not change over time. Widespread methods of biometric identification, such as fingerprinting, retina recognition, and voice recognition have various vulnerabilities unfortunately.

Keywords: biometric authentication, electrocardiogram, information security

1. Fingerprinting

This identification method is widely used by the FBI.

The FBI has managed the nation's collection of fingerprints since 1924, but it went fully electronic in 1999 when launched the Integrated Automated Fingerprint Identification System, or IAFIS. This national repository of fingerprints and criminal histories enables law enforcement at every level to quickly match up criminal evidence with criminal identities [1].

On the other hand, the Department of Homeland Security's IDENT—the Automated Biometric Identification System that houses fingerprint records and limited biographic information—was created in 1994 to help U.S. border and immigration officials keep criminals and terrorists from crossing US borders.

In this post-9/11, globalized world, the Department of Justice (DOJ) and FBI, the Department of Homeland Security (DHS), and the Department of State have worked hard in recent years to establish interoperability between these two fingerprint databases.

2. Retina recognition

Retina recognition is a biometric technique that uses the unique patterns on a person's retina for person identification. The retina is the layer of blood vessels situated at the back of an eye. The eye is positioned in front of the system at a capture distance ranging from 8 cm to 1 m. The person must look at a series of markers, viewed through the eyepiece, and line them up. The eye is optically focused for the scanner to capture the retina pattern. The retina is scanned with the near infrared (NIR 890 nm) irradiation, and the unique pattern of the blood vessels is captured. Retina recognition makes use of the individuality of the patterns of the blood vessels. It has been developed commercially since the mid-1970s. Sandia Laboratory reported a false rejection rate of lower than 1.0% [2].

3. Voice recognition

Voice biometrics is the science of using a person's voice as a uniquely identifying biological characteristic in order to authenticate them. Also referred to as voice verification or speaker recognition, voice biometrics enables fast, frictionless, and highly secure access for a range of use cases from call center, mobile, and online applications to chatbots, IoT devices, and physical access.

Like other biometric modalities, voice offers significant security advantages over authentication methods that are based on something you know (like a password or answer to a "secret" question) or something you have (like your mobile phone). Voice biometrics also improves the customer experience by removing frustration associated cumbersome login processes and lost and stolen credentials [3].

Many banks use the voice recognition technologies for person identification.

Unfortunately, the above biometric identification technologies are not free from some disadvantages. In particular, there are methods to fake fingerprints [4], retina [5], and voice [6].

Currently, biometric identification technologies such as ECG, EEG, and DNA are considered resistant to hacking.

In this chapter, we would like to talk about biometric identification using ECG and related problems.

4. What is an ECG?

There are several scenarios for using biometric identification using ECG.

1. Contact ECG

2. Remote ECG sensing

For contact ECG, you can use medical electrocardiographs or sensors installed in smartphones or, for example, in the steering wheel of a car. An interesting direction in the development of computer technology is wireless sensor networks. A special case of this technology is Body Area Networks (BAN), which include one or more sensors that record the medical parameters of the human body. BANs are Internet of Medical Things (IoMT). ECG sensors in this case are mounted on the patient's body. Medical parameters are transmitted to the server for analysis and storage. An ECG can identify the patient.

Remote ECG sensing is carried out using ultra-wideband radars. An example is the instrument of the Israeli military company XAVER, which allows special forces to detect living people through a brick wall, as well as determine their location, gender, and age (**Figure 1**) [7].

More advanced is the technology of WAVD Technology, Arizona. Its ultra-wideband radar allows not only to detect, but also to identify people buried under the rubble of buildings.

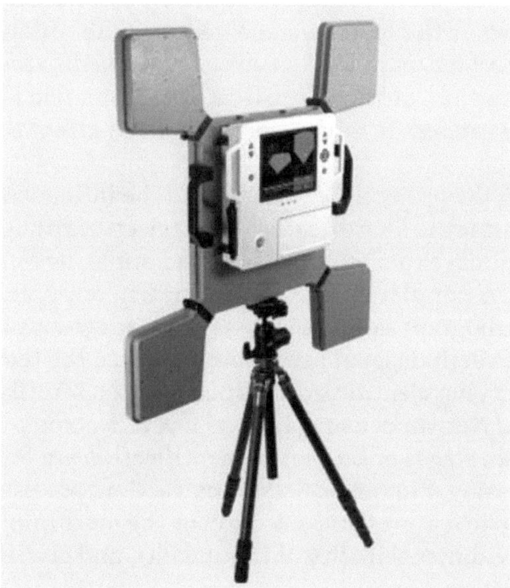

Figure 1.
Xaver 800. Wall-through 3D imagine system.

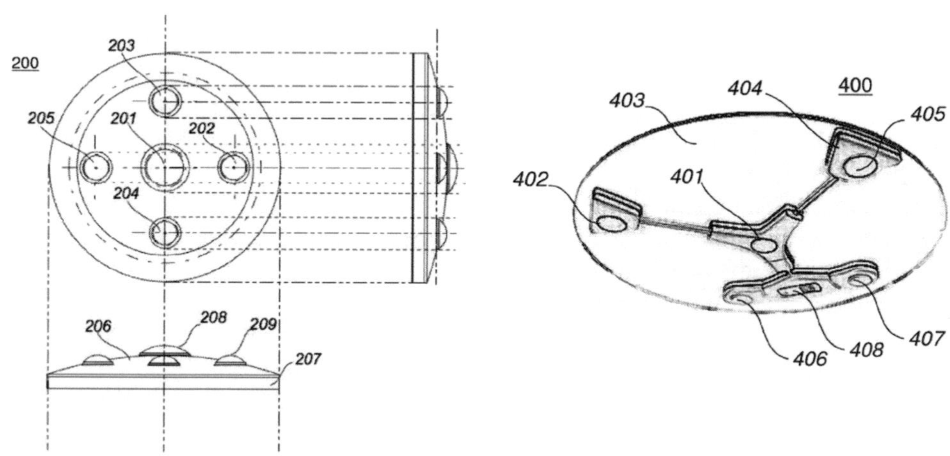

Figure 2.
The use of ultra-wideband radars for biometric identification. Left: banking sector; right: security systems.

Several patents have been published that propose the use of ultra-wideband radars for conducting banking operations without a credit card, as well as for controlling premises during confidential meetings (**Figure 2**). As it turned out, with the help of ultra-wideband radars, it is possible to restore not only the ECG, but also speech.

Xiaolin et al. described using of ultra wide band radars for detecting of vital signs [8, 9].

5. ECG-based biometric identification structure

Biometric identification involves the stage of user registration and the stage of user recognition. At the registration stage, the user takes biometric features and writes them to the database. At the recognition stage, biometric features are taken from an unknown person and consequently compared with the features stored in the database. If the features received from an unknown person by a certain criterion coincided with the features from the database, then a decision is made on the success of the identification. Biometric identification is a complex multi-stage process in which each stage can affect the final recognition accuracy.

While performing the project, we investigated the influence of various factors on the accuracy of biometric identification using electrocardiograms. To do this, a large-scale computational experiment was carried out using our programs written in Python. We used the popular libraries such sklearn, scipy, and matplotlib. Most digitalized electrocardiogram samples were taken from www.physionet.org website. When performing the digital signal processing, we used the biosppy and wfdb libraries. When classifying electrocardiograms, we used Multilayer Perceptron and Convolutional Neural Networks using TensorFlow technology.

The following main stages of biometric identification are follows: signal registration, signal preprocessing, biometric feature extraction, assessment of the informativeness of biometric features and selection of the most informative features (this is done to reduce dimensionality of input data), and classification of features. Consider each of the steps.

6. Registration of an electrocardiogram

Electrocardiogram (ECG or EKG [a]) is a graph of voltage versus time – of the electrical activity of the heart using electrodes placed on the skin. These electrodes detect the small electrical changes that are a consequence of cardiac muscle depolarization followed by repolarization during each cardiac cycle (heartbeat) (**Figure 3**).

From a technical point of view, Electrocardiographs are multichannel voltmeters that record electrical potentials in various areas of human surface. These devices differ in such characteristics as sampling frequency, bit depth, input voltage range, etc. A valuable resource for researchers in the field of analysis of biomedical signals is the website https://www.physionet.org/. PhysioNet is a repository of freely available medical research data, managed by the MIT Laboratory for Computational Physiology. The project is supported by the National Institute of General Medical Sciences (NIGMS) and the National Institute of Biomedical Imaging and Bioengineering (NIBIB) under NIH grant number 2R01GM104987-09.

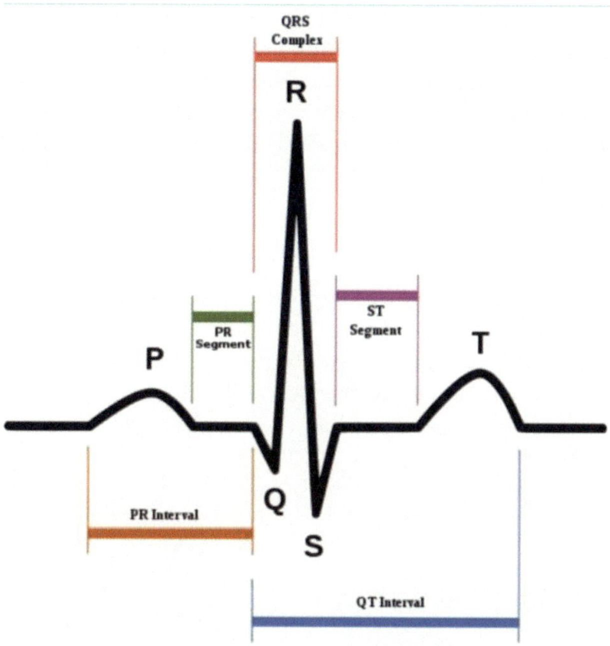

Figure 3.
ECG of a heart in normal sinus rhythm (https://en.wikipedia.org/wiki/Electrocardiography).

This site presents a large number of digitized electrocardiograms, for example, PTB Diagnostic ECG Database [10]

- 16 input channels,
- Input voltage: ±16 mV,
- Input resistance: 100 Ω (DC),
- Resolution: 16 bit with 0.5 μV/LSB (2000 A/D units per mV),
- Bandwidth: 0–1 kHz (synchronous sampling of all channels, time of registration is 3 minutes).

European ST-T Database [11]

- each record is two hours in duration and contains two signals,
- each sampled at 250 samples per second with 12-bit resolution over a nominal 20 millivolt input range.

ECG-ID Database [12, 13]

- ECG lead I, recorded for 20 s,
- digitized at 500 Hz with 12-bit resolution over a nominal ±10 mV range

As we can see, all the above databases used different electrocardiographs.

Previously, we investigated factors that influence the accuracy of biometric identification using an ECG. We have shown that the quality of an electrocardiograph affects the accuracy of biometric identification. Thus, the recognition accuracy during ECG classification using mixed Gaussian models of subjects from the ECG-ID database was 0.66, while for PTB this indicator was 0.8 [14].

7. Signal preprocessing

Signal preprocessing is carried out in order to reduce noise, reduce data dimension, find R-peaks, and cut ECG into cardiocycles. In this case, R-peak synchronization is usually performed. Noises are usually removed using a low-pass filter, while the cutoff frequency is selected experimentally. There is still no consensus on how best to find R-peaks. This is due to the fact that cardiocycles in different people are

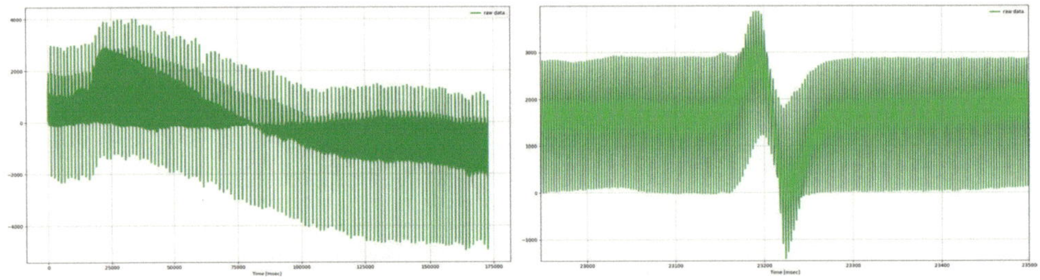

Figure 4.
The original digitized electrocardiogram has a high frequency.

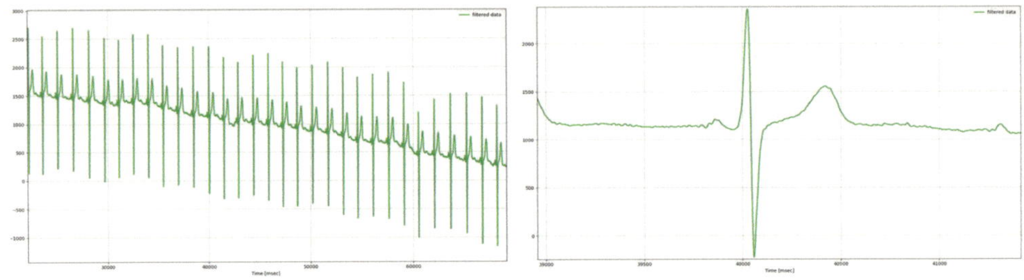

Figure 5.
Using the low-pass filter, the envelope of the cardiac signal is extracted.

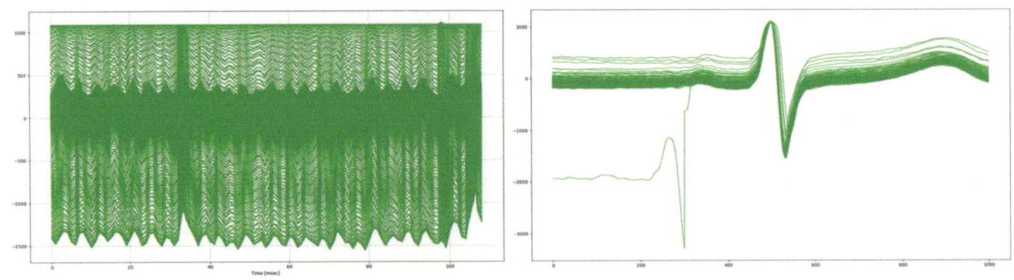

Figure 6.
Further, R-peaks are detected in the ECG, with their help the signal is cut into cardiocycles, after which the R-peak is synchronized.

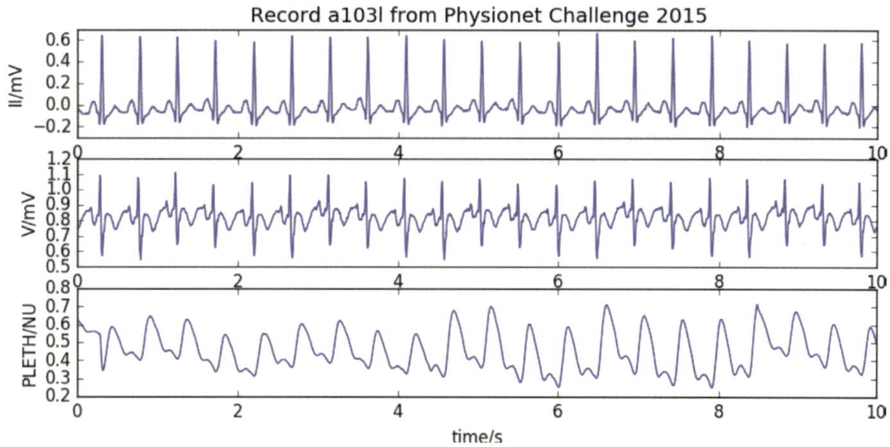

Figure 7.
Wfdb library example.

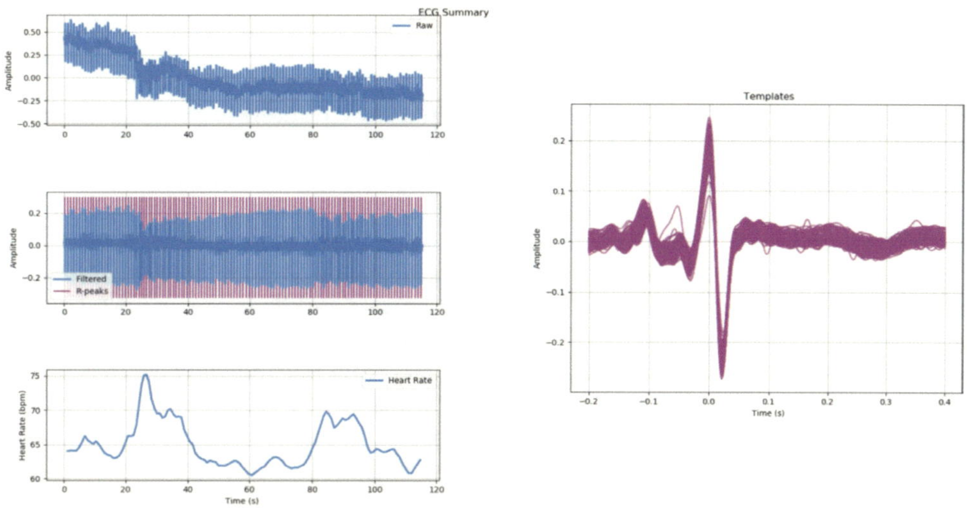

Figure 8.
Example of using the biosppy library.

distinguished by a rather high degree of great variability. In **Figure 4**, a digitized electrocardiogram is shown (sample rate is 1000 Hz).

To reduce noises, we can use a low-pass filter (**Figure 5**).

After noise removal, the procedure for finding R-peaks, slicing an ECG into cardiocycles and synchronizing cardiocycles by R-peaks follows (**Figure 6**).

There are several popular computer libraries for ECG preprocessing. Among them, libraries for the python wfdb [15] and biosppy [16] languages are very popular (**Figures 7** and **8**).

8. Biometric features extraction

There are several opinions as to which ECG features are best used for biometric identification. Some authors propose using the geometric characteristics of the cardiocycle, such as the amplitude and time characteristics of the cardiocycle peaks.

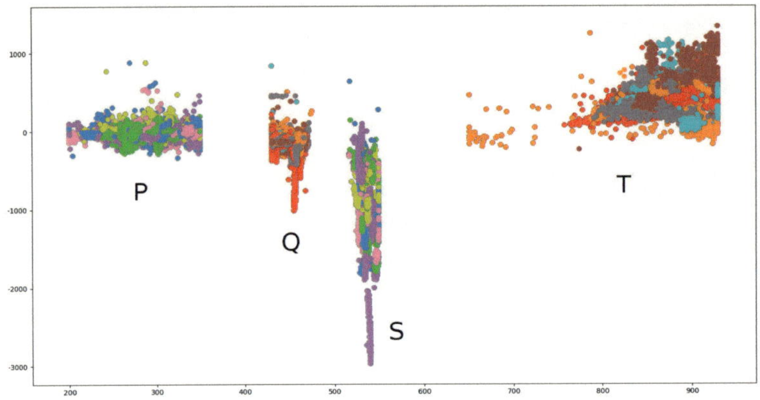

Figure 9.
Amplitude and time characteristics of P, Q, S, T peaks. Cardiocycles are synchronized in amplitude and time of onset of R-peak.

We can see a feature cloud in **Figure 9**, which are the amplitude and time characteristics of P, Q, S and T peaks of cardiac cycles.

Other authors suggest working with the frequency characteristics of the signal. For example, biometric features can be obtained using a discrete wavelet transform. Previously, we explored wavelets such as Haar wavelets, Daubechies wavelets (from db1 to db38), Symlets (from sym2 to sym20), Coiflets (from coif1 to coef17), Biorthogonal (from bior1.1 to bior6.8), Reverse biorthogonal wavelet (from rbio1.1 to rbio6.8), and Discrete Meyer (FIR Approximation) [17]. We have shown that wavelets such as Haar, Daubechies, and Symlets are best suited for biometric identification. We have shown that good results can be obtained if the entire cardiocycle is used as biometric features [18]. The number of features in this case depends on the sampling frequency of the signal. We used data from the following databases. PTB database (sampling rate is 1000 Hz), the cardiocycles consist of 600 points, in the case of the European ST-T Database (sampling rate is 250 Hz), the cardiocycles consist of 150 points, and in the case of St.Petersburg Institute of Cardiological Technics 12-lead Arrhythmia Database (sampling rate is 257 Hz), cardiocycles consist of 153 points.

9. Assessment of the informative value of biometric features and the selection of the most informative features

Experience shows that not all biometric features have the same information content. If you remove of uninformative features, you can significantly increase the speed of data processing. We investigated the informativeness of analytical features (amplitude and time characteristics of P, Q, S, T peaks) obtained from 51 subjects from the PTB database [19]. To do this, we determined the significance of differences between the clouds of points P, Q, S and T regions of the electrocardiograms of the subjects using Student's criterion at a significance level of 95%. Matrices of significance of differences are given below (**Figure 10**).

It can be seen from the figure that the overlap of the points is much smaller in the S and T regions. When using all eight signs together, the overlap of the points is not observed (**Figure 11**).

Conclusion: the most informative analytical features are the amplitude values in the S and T regions.

DOI: http://dx.doi.org/10.5772/intechopen.91172

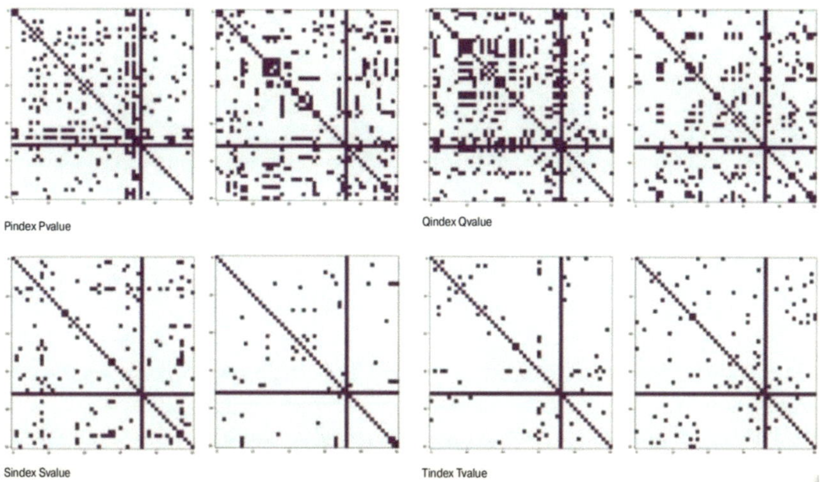

Figure 10.
Matrices of significance of differences according to 8 characteristics for 51 subjects (P value < 0.05). Note: in the bright areas of the figures, the differences are significant, in the dark areas – unreliable. The figures are symmetrical with respect to the diagonals passing through the upper left and lower right corners. The abscissa and ordinate axes show the numbers of subjects (1–51).

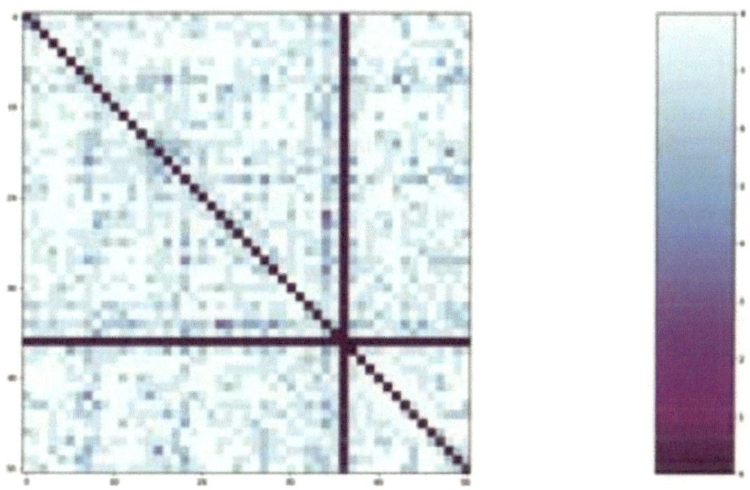

Figure 11.
Matrix of significance of differences when sharing eight features. Note: the color shows the number of cases of significance of differences from eight (the lightest area) to zero (the darkest area). The pattern is symmetrical with respect to the diagonal passing through the upper left and lower right corners. The abscissa and ordinate axes show the numbers of subjects (1–51).

10. Feature classification

The problem of person biometric identification concerns classification problems. To solve it, we have to consider algorithms from some finite set and choose an algorithm that gives the least error of the forecast. Let's introduce some notation. Let us suppose X is a space of objects.

Y is a set of answers.

$$X^l = (xi, yi)^l{}_{i=1} \quad (1)$$

is a training set, l is a sample size.

$$y_i = y^*(x_i), \tag{2}$$

$$A_t = \{a:X \to Y\} \tag{3}$$

are a model of algorithms, t ⊆ T, T is a number of algorithms under consideration.

$$\mu t:(X \times Y)^l \to A_t \tag{4}$$

are learning methods. It is required to find a method μ t with the best generalizing power.

When finding a method μ_t, we often have to solve the following subtasks:

- Choice of the best model A_t (model selection).
- Choice of learning method μ_t for a given model A_t (in particular, optimization of hyperparameters).
- Features selection:

$$F = \{f_j:X \to Dj:j = 1,\dots,n\} \tag{5}$$

is a set of features. The method of learning μ j uses only features J ⊆ F.

It is used to assess the quality of learning by precedents.

$L(a, x)$ is a cost function of algorithm a on the object x.

$$Q(a, X^l) = \frac{1}{l}\sum_{i=1}^{l} L(a, x_i) \tag{6}$$

is a functional of accuracy a on X. In this case, we consider an internal quality criterion that is measured on the training set X^l:

$$Q_\mu(X^l) = Q(\mu(X^l), X^l) \tag{7}$$

and an external criterion evaluating the quality of learning on hold-out set Xk [2]:

$$Q_\mu(X^l, X^k) = Q(\mu(X^l), X^k) \tag{8}$$

Recognition accuracy will be affected by both the choice of the classification method and its implementation, in particular, the selection of hyperparameters. We tested 14 methods of Machine Learning for classification (Naive Bayes classifier for multivariate Bernoulli models, A decision tree classifier, An extremely randomized tree classifier, Classifier implementing the k-nearest neighbors vote, Label Propagation classifier, Linear Discriminant Analysis, Linear Support Vector Classification, Logistic Regression (aka logit, MaxEnt) classifier, Nearest centroid classifier, A random forest classifier, Classifier using Ridge regression, Ridge classifier with built-in cross-validation, and Gaussian Mixture Models, SVM) [20]. We found that the most accurate methods of classification are Label Propagation classifier (accuracy of recognition is 0.94), an extremely randomized tree classifier (accuracy is 0.92), and a Classifier implementing the k-nearest neighbors vote (accuracy is 0.90) [21].

By selecting model hyperparameters, it is possible to significantly increase recognition accuracy. So, in our previous study, it was shown that using the Support Vector Machine classifier for ECG classification uses as default following hyper parameters: C = 1.0, kernel = "rbf," gamma = "auto." When using of default parameters while performing of classification of electrocardiograms, we had an accuracy score equal to 0.93. We tuned hyper parameters of classification with Grid Search procedure varying C parameter in range of [1, 10, 100, 1000], kernel in range of ["linear," "rbf"], and gamma in range of [1e-3, 1e-4]. After performing of tuning, we had the following best parameters set: "kernel": "rbf," "C": 10, "gamma": 0.001. Using these parameters, we had an accuracy score equal to 0.99.

11. Conclusion and perspectives

Traditional password-based authentication methods have a number of disadvantages related primarily to the human factor. Biometric methods of identification and authentication are much more reliable, although they have some disadvantages. Some of them (fingerprints, retina, and voice) were compromised. It is not clear what to do if hackers gain access to a biometric database, because a person cannot change fingerprints as easily as a forgotten password. The development of wireless technologies and technologies of the Internet of medical things makes possible the emergence of new biometric identification scenarios. Here, first of all, I would like to note the biometric authentication of the patient in the Body area network. In this case, ECGs are not used to generate biometric keys confirming the patient's authenticity [22]. The second important area is contactless ECG recording using ultra-wideband radars.

Acknowledgements

The reported study was funded by RFBR according to the research project no. 19-07-00780.

Author details

M.R. Bogdanov[1,2*], A.S. Filippova[2], G.R. Shakhmametova[1] and Nikolai N. Oskin[3]

1 Ufa State Aviation Technical University, Ufa City, Russia

2 M.Akmullah Bashkir State Pedagogical University, Ufa City, Russia

2 Siberian Telecommunication Company, Moscow, Russia

*Address all correspondence to: bogdanov_marat@mail.ru

References

[1] Fingerprint Technology. Making Two Systems Work as One. The web site of Federal bureau of Investigation [Internet]. Available from: https://archives.fbi.gov/archives/news/stories/2010/july/fingerprints [Accessed: 07 January 2010]

[2] Seto Y. Retina recognition. In: Li SZ, Jain A, editors. Encyclopedia of Biometrics. Boston, MA.: Springer; 2009

[3] What is Voice Biometrics and why should you use it? The web-site of ID R&D Co. [Internet]. Available from: https://www.idrnd.ai/voice-biometrics/

[4] Software for fingerprint spoof and liveness detection. The web-site of Precise Biometrics Co. [Internet]. Available from: https://precisebiometrics.com/products/fingerprint-spoof-liveness-detection/software/

[5] Lai CL, Tai CY. A smart spoofing face detector by display features analysis. Sensors. 2016;**16**:1136. DOI: 10.3390/s16071136

[6] Evans N, Kinnunen T, Yamagishi J, Wu Z, Alegre F, De Leon P. Anti-spoofing for speaker recognition. In: Marcel S, Nixon MS, Li SZ, editors. Handbook of Biometric Anti-Spoofing. Trusted Biometrics under Spoofing Attacks; 2014

[7] XAVER™ 800. High performance imaging system. The web-site of Camero, part of the SK Group. Available from: https://www.camero-tech.com/xaver-products/xaver-800/

[8] Xiaolin L, Jianqin D, Hao Z, Thomas AG. Ultra-wideband impulse radar through-wall detection of vital signs. Scientific Reports. 2018;**8**:13367. DOI: 10.1038/s41598-018-31669-y

[9] Chocorresponding H-S, Park Y-J. Detection of heart rate through a wall using UWB impulse radar. Journal of Healthcare Engineering. 2018;**2018**:4832605. DOI: 10.1155/2018/4832605

[10] Bousseljot R, Kreiseler D, Schnabel A. Nutzung der EKG-Signaldatenbank CARDIODAT der PTB über das Internet. Biomedizinische Technik. 1995;**40**(1):317

[11] Taddei A, Distante G, Emdin M, Pisani P, Moody GB, Zeelenberg C, et al. The European ST-T database: Standard for evaluating systems for the analysis of ST-T changes in ambulatory electrocardiography. European Heart Journal. 1992;**13**:1164-1172

[12] Lugovaya TS. Biometric human identification based on electrocardiogram [Master's thesis]. Faculty of Computing Technologies and Informatics, Electrotechnical University "LETI", Saint-Petersburg, Russian Federation; 2005

[13] Goldberger AL, Amaral LAN, Glass L, Hausdorff JM, Ivanov PC, Mark RG, et al. PhysioBank, PhysioToolkit, and PhysioNet: Components of a new research resource for complex physiologic signals. Circulation. 2003;**101**(23):e215-e220

[14] Bogdanov MP et al. Factors influencing accuracy of biometrical personal identification based on cardiograms. Pattern Recognition and Image Analysis. 2018;**28**(3):421-426

[15] The WFDB Python Toolbox [Internet]. Available from: https://pypi.org/project/wfdb/

[16] BioSPPy is a toolbox for biosignal processing written in Python [Internet]. Available from: https://biosppy.readthedocs.io/en/stable/

[17] Bogdanov M et al. Processing of biomedical data with machine learning. Atlantis highlights in computer sciences.

21st International Scientific Workshop on Computer Science and Information Technologies (CSIT 2019). 2019;**3**:6-16

[18] Bogdanov M et al. Increasing security of telemedicine service. Atlantis highlights in computer sciences. 21st International Scientific Workshop on Computer Science and Information Technologies (CSIT 2019). 2019;**3**:162-165

[19] Bogdanov MP et al. Statistical assessment of informativeness of biometric features extracted from electrocardiograms. Russian Cardiological Journal. 2018;**23**(7):84-91

[20] Bogdanov MR et al. Diagnosis of heart diseases with machine learning. Journal of Mathematics and Statistical Science. 2019;**5**:81-84

[21] Bogdanov MR et al. Optimizing Factors Influencing on Accuracy of Biometrical Cardiometry. In: Belim S, editor. Biometrical Cardiometry. Omsk, Russia, published at http://ceur-ws.org: OPTA-SCL; 2018. pp. 61-64

[22] Nima K, Zimu G, Fatemeh T, Damon W, Mark T, Domenic F. Secure and Reliable Biometric Access Control for Resource-Constrained Systems and IoT. Available from: https://www.researchgate.net/publication/324055574_Secure_and_Reliable_Biometric_Access_Control_for_Resource-Constrained_Systems_and_IoT

Chapter 3

Face Identification Using LBP-Based Improved Directional Wavelet Transform

Mohd. Abdul Muqeet and Qazi Mateenuddin Hameeduddin

Abstract

Face identification is the most active area of research in computer vision and biometric authentication. Various face identification methods are developed over the time, still, numerous facial appearances are needed to cope with such as facial expression, pose, and illumination variation. Moreover, faces captured in unrestrained situations also impose immense concern in designing effective face identification methods. It is desirable to extract robust local descriptive features to effectively characterize such facial variations both in unrestrained and restrained situations. This chapter discusses such a face identification method that incorporate a popular local descriptor such as local binary patterns (LBP) based on the improved directional wavelet transform (IDW) method to extract facial features. This designed method is applied to complex face databases such as CASIA-WebFace and LFW which consists of a large number of face images collected under an unrestrained environment with extreme facial variations in expression, pose, and illumination. Experiments and comparison with various methods which include not only the local descriptive methods but also local descriptive-based multiresolution analysis (MRA) based methods demonstrate the efficacy of the LBP-based IDW method.

Keywords: face identification, improved directional wavelet (IDW), local binary patterns (LBP)

1. Introduction

Researchers have devoted a substantial amount of effort in studying face identification methods in the context of computer vision, image processing, and pattern recognition due the wide acceptability of face as biometric [1]. Requirements for high recognition accuracy, high computational efficiency, and invariance to variations in facial expression, illumination, pose, and occlusions are the prominent challenges in face identification. The illumination problem comes from the fact that different illuminations can cause vast changes to the image of a subject's face [2, 3]. Similarly, deviations in facial expressions along with head pose variation and occlusion can also lead to very unlike face images for the same subject. Moreover, face identification in the unrestrained environment is still a major challenge which greatly degrades the performance of various well-established methods. Additionally, there still exist challenges such as high dimensionality of feature data and intraclass variations occurring due to the effect of facial variations in restrained and unrestrained environments. A face identification method must be discriminatory

IntechOpen

for different subjects and invariant to numerous facial variations. Researchers have been extensively utilizing MRA methods and using various off–the-shelve designs of wavelet filters [4] for the implementation of isotropic 2-D DWT for facial feature extraction. Recently implemented 2-D DWT methods such as GHWFB [5] and TWFB [6] considers the handcrafted wavelet filters with additional features compared to off-the-shelve wavelet filters. But these methods do not achieve excellent results due to limited directions orientation and non-adaptation in facial feature selection.

Various local descriptors prominently the LBP [7] and weber local descriptors (WLD) [8] have been efficiently used for facial feature extraction. The constraint of the LBP-based feature extraction method is their noise intolerance and poor discrimination capability [8]. Recently, various non-adaptive MRA methods are applied as a pre-processing step before LBP feature extraction to improve the performance. The prominent methods are local Gabor binary patterns (LGBP) [9], Steerable Pyramid Transform (SPT)-LBP [10], Curvelet Transform-LBP (Curvelet-LBP) [11], Contourlet-LBP [12], and Wavelet Transform (WT) LBP [13]. Liu et al. [14] used hierarchical multi-scale LBP to create features of sparser coefficients and performed classification using sparse coding with the application of a greedy search approach. Wang et al. [15] combined the Gabor wavelet transform (GWT) and CLBP features and carried out the SRC to perform classification.

The aforementioned LBP-based MRA methods [9–13, 15] use non-adaptive directional transform which lacks the adaptive directional selectivity based on the image description. These methods also experience various issues, for instance, selection of transformed sub-bands, complex filter design, and the large dimension of the feature vector. Maleki et al. [16] proposed adaptive direction selection and applied directional lifting within the selected optimal direction and constructed a compact representation for adaptive MRA method. Due to such inherent characteristics, significant directional details for various face variations can be approximated by the detection of edges responsible for such variations [17].

For numerous facial variations, substantial directional details can be estimated by approximating the edges [18, 19] accountable for such variations which will considerably enhance the face identification performance which decides the basis of our method. The concept has been exploited in [17–19] for face recognition applications. This work extends the design of the adaptive directional scheme presented in [19] and presents an LBP-based IDW method to capture multi-resolution directional details from the face images. Subsequently in contrast to [19] where CLBP is used, LBP is applied to the generated IDW sub-bands to extract MRA-based local descriptive features.

The Implementation of the 2-D IDW using seven directions along with the quadtree partitioning (QTP) scheme [19] is explained in Section 2. A brief theory on LBP is described in Section 3. Further, the proposed facial feature extraction method is exhibited in Section 4. In Section 5, comparative results on the CASIA-WebFace and LFW face databases are demonstrated. Conclusions are highlighted in Section 6.

2. Implementation of 2-D improved directional wavelet (2-D IDW)

The fundamental concept of implementation of improved directional wavelet (IDW) is to carry out transform operations on a face image at a viable variety of possible directions while maintaining the properties of multi-resolution, localization, and isotropy intact. The authors in [19] considered a set of seven directions with a quad-tree partitioning scheme. Here we will provide a brief review of the work mentioned in [19].

The 2-D IDW being isotropic method performs a separable 1-D horizontal and vertical 1-D IDW on face image with variation in prediction and update steps where seven directions are considered in an adaptive direction scheme. While performing the 1-D IDW transform if non-integer sample arrives sub-pixel interpolation is performed.

Let $x[i, j]$ be a 2-D face image which is first horizontally sub-sampled to get even subsamples $x_e[i, j] = x[2i + j]$ and odd sub-samples $x_o[i, j] = x[2i + 1, j]$. Next in the prediction step odd samples $x_o[i, j]$ are predicted from neighboring even samples with strong correlation along an optimal direction θ. The outcome of the prediction step and the generated high-pass signal $H[i, j]$ are described as [19],

$$P(x^o)[i, j] = \sum_{n=-N_p}^{N_p-1} K_n^p . x_e[i + n, j + \text{sign}(n - 1)\tan\theta] \tag{1}$$

$$H[i, j] = g_H(x_e[i, j] - P(x^o)[i, j]) \tag{2}$$

Where K^p and $2N_p$ are the length and coefficients of the prediction filter. Here, samples from six even rows are selected to conduct the prediction step [19]. Now in the update step, odd samples of $H[i, j]$ along the same optimal direction θ as used in (1) are selected to modify the even samples. The update step and the generated low-pass signal $L(i, j)$ are given as [19],

$$U(H)[i, j] = \sum_{n=-N_u}^{N_u-1} K_n^u . \begin{pmatrix} x_o[i + n, j + \text{sign}(n)\tan\theta] - \\ P(x_o)[i + n, j + \text{sign}(n)\tan\theta] \end{pmatrix} \tag{3}$$

$$L[i, j] = g_L\left(x_o[i, j] + g_H^{-1}(U(H_o)[i, j])\right) \tag{4}$$

Where K^u and $2N_u$ are the coefficients and length of the update filter. Similarly, samples from six odd rows are selected to conduct the update step. The values of scaling factors are considered as $g_L = 1.3416$ and $g_H = 0.7071$ [17, 19].

Due to linear phase characteristics and large vanishing moments, Neville filters with the order as six are considered as the coefficients K^p and K^u [19]. The usage of Neville filters increases the approximation ability of 2-D IDW.

In contrast to the nine directions [17] and five directions [18], we also used seven pre-assigned directions to implement 2-D IDW [19].

$$\Theta = \{\theta | \theta = 0, 22.5, 45, 67.5, 90, 112.5, 135\} \tag{5}$$

These directions are used to confirm a strong correlation among samples and to extract directional MRA features from face images. It is to point out that $\text{sign}(n - 1)\tan\theta$ term in (1) and (3) may not always locate integer samples and may not be present on the original image sampling grid [19]. So; sub-sample interpolation is conceded to compute intensity for such non-integer samples. To maintain perfect reconstruction lifting structure [4], the integer samples required to perform sub-sample interpolation for such non-integer samples at optimal direction θ must be even sampled. If optimal direction comes across the integer samples the value is computed by the nearby even sample otherwise the value of the non-integer sample is computed from the interpolation of the two nearby even samples.

To extract local edge details due to face variations that exist at different pixel regions, a quadtree partitioning (QTP) scheme is implemented to partition each face image into sub-blocks of distinct directional details. Each QTP sub-block will have the same direction. The improved QTP scheme provides an efficient direction assignment while implementing the prediction and update step.

Let each face image $x[i, j]$ is applied with QTP to obtain non-overlapping sub-block x_s. Also, consider the initial block size S_{ini}, minimum block size as S_{min} and the Lagrangian multiplier as α. The energy summation of the prediction error (ESPE) for each block is computed as [19],

$$ESPE_{s,n} = \sum_{i,j \in R_{s,n}} \|x_s[i,j] - F_{s,n}[i,j]\|_2^2 + \alpha B^n \tag{6}$$

Where $F_{s,n}[i, j]$ are the filtered responses obtained by applying the prediction filter K^p along with the predefined directions θ. B^n is the number of bits spent on signaling the selection of directions. When a sample is predicted from the nearest samples, each candidate direction from (5) is checked and the direction with the smallest ESPE is ultimately selected. The optimal direction which gives the least value of ESPE is selected as,

$$\theta_s = \arg\min_n \{ESPE_{s,n}\} \tag{7}$$

The value of the lagrangian multiplier α determines the complexity of the QTP scheme and its value needs to be selected sensibly. Moreover, to detect the local edge details and to suit it to the adaptability of the IDW method, a face image needs to be segmented into partitions of clear orientation bias. To resolve this problem an improved QP scheme is proposed to suit the face identification problem as mentioned in [19]. The 1-D IDW can be simply extended to the 2-D IDW where second dimension lifting is yet again performed in the horizontal direction on high-pass signal $H[i, j]$ and low-pass signals $L[i, j]$ to generate four sub-bands i.e. $LH[i, j]$, $LL[i, j]$, $HH[i, j]$, and $HL[i, j]$.

3. Local binary patterns

The LBP [7] is estimated with sampling points $x_p \in (0, \ldots, P-1)$ in the neighborhood of a center pixel $x_m(i_c, j_c)$ at a radial distance given by R [7],

$$LBP_{P,R} = \sum_{p=0}^{p-1} t_s.(x_p - x_m).2^p \tag{8}$$

$$t_s(d) = \begin{cases} 1, (d) \geq 1 \\ 0, (d) < 1 \end{cases} \tag{9}$$

Where $t_s(d)$ is a threshold function. The sampling points which does not fit within the center of a pixel are bilinearly interpolated [7]. Another extension of LBP is the uniform patterns and it is mapped from $LBP_{P,R}$ to $LBP^{u2}_{P,R}$ [18], resulting in $P \times (P-1) + 3$ feature dimension. After obtaining the LBP coded image, codes of the input image $X_L(i, j)$ pixels are formed into a histogram as a feature descriptor,

$$H_l = \sum_{i,j} F\{X_L(i,j) = l\}, F\{y\} = \begin{cases} 1, \text{if y is true} \\ 0, \text{if y is false} \end{cases}, l = 0, 1, 2 \ldots, n-1 \tag{10}$$

Where n is the number of different labels produced by the LBP operator. With the usage of $LBP^{u2}_{8,1}$, the feature dimension is 59 [18].

4. Implementation of LBP-based IDW method

We consider a resolution of 128 × 128 pixels for face images of the selected databases and face preprocessing is performed on all the face images. Thus each one of the IDW sub-bands {LL, HL, LH} is of size 32 × 32 pixels and each sub-band is divided into m = 16 regions with the size of each region as x × y = 8 × 8 pixels [18]. We applied LBP to each of the regions from each of the sub-bands {LL, HL, LH}. We used the uniform pattern $LBP^{u2}_{8,1}$ [7] and NN classifier with Chi-square distance measure. This form an enhanced feature vector or descriptor *EFV* with a combined dimension as 59 × m × 3 = 2832 [18]. The algorithm representing the proposed method is presented in Algorithm 1.

Algorithm 1: Face Identification using LBP-based IDW

Input: Test Image, Train image
Output: Rank-one recognition results of the feature vectors.
Algorithm:
Step 1: (Preprocessing)
1.1. Consider the input face image X.
1.2. Resize the image to the resolution of 128 × 128 pixels.
Step 2: (Computation of IDW sub-bands)
for a number of decomposition levels **do**
2.1. Quadtree partitioned the face image X into several non-overlapping sub-blocks.
2.2. Estimate the value of ESPE using (6) for each sub-block.
2.3. Estimate the optimal direction which gives the least value of ESPE using (7).
2.4. Perform the prediction and the update steps as described in (1) and (3) in the selected directions in the selected sub-block.
2.5. Obtain IDW sub-bands {LL, HL, LH, HH} and proceed with the LL sub-band for the next decomposition level.
Step 3: (LBP Computation)
3.1. Consider the top-level {LL, LH, HL} sub-bands and divide each sub-band into non-overlapping regions R_k with each of size 8 × 8 pixels.
for each sub-band **do**
for each sub-block within the sub-band **do**
for each coefficient value within the sub-block **do**
3.2. Compute the $LBP^{u2}_{8,1}$ histogram features from each region R_k using (8), (9), and (10).
3.3. Concatenate all such $LBP^{u2}_{8,1}$ multi-region histograms from each sub-band {LL, HL, LH} to form the histogram feature vectors $LL_{l,k}$, $HL_{l,k}$, and $LH_{l,k}$ respectively.
end for end for end for
3.4. Concatenate all the sub-band histogram features to form the final enhanced histogram feature vector *EFV*.
Step 4: (Dimensionality Reduction)
4.1. Perform dimensionality reduction using LDA on the *EFV* feature vector.
4.2. Save the reduced dimension train feature vector database to *EFV_train*.
Step 5: Repeat Step 1 to 3 and 4.1 on each test image to obtain the test feature vector *EFV_test*.
Step 6: (Identification)
6.1. Compare test feature vector *EFV_test* against train feature vector *EFV_train* using the NN classifier using Chi-Square distance measure and calculate the Rank-one results in an identification process.

5. Experimental results

All experiments are performed using Matlab 2018a on a standard i5–3320 2.60 GHz machine with 8.0 GB RAM. Here we prove the effectiveness of the LBP based IDW feature extraction method for which a comparison is established with other LBP based non-adaptive MRA methods. In the experiments, we randomly select a few face images for training and rest for testing to obtain the recognition results.

The face identification performance of the proposed method is performed on CASIA-WebFace [20] and LFW [21] face Database with extreme facial variations where all the images are considered under unrestrained environment. Since, LBP-based IDW histogram features are extracted, comparative face identification methods include descriptive methods such as LBP [7], WLD [8], and SRC-GSLBP [14]. Besides, a comparison with few non-adaptive MRA-based LBP feature extraction methods such as LGBPHS [9], SPT-LBP [10], Curvelet-LBP [11], Contourlet-LBP [12], and GTCLBPSRC [15] is also performed for competitive analysis.

5.1 Experiments on the CASIA-WebFace face database

The CASIA-WebFace database [22] is a huge and complex face database. This database includes 494,414 face images of 10,575 subjects. Considering the subjects with only a few samples deters the recognition results. Thus 10,575 subjects are allocated in the decreasing order by the count of their images contained in the particular subject set. Here we consider only 9067 subjects which consist of at least 15 images. The remaining images of the rest of the 1508 subjects are discarded. Within this, we considered a subset of 600 subjects with 15 images per subject out of 9067 subjects. These subjects are specifically considered based upon their extreme facial variations. Face images are normalized and resized to128x128 pixels.

A random subset is constructed with T (T = 4, 5, 6, 7, 8) images of each subject for training and in every case remaining images for testing. **Table 1** tabulates the Rank-one recognition results of various comparative methods. Since the images are collected from around the web with extremely unrestrained conditions, the results are less for all the methods. Due to extreme face variations which include

Method	Number of training samples per subject				
	4	5	6	7	8
LBP [7]	14.07	16.06	18.24	21.81	24.14
WLD [8]	15.60	17.71	21.40	23.20	25.09
LGBPHS [9]	22.80	26.09	28.93	31.48	34.76
SPT-LBP [10]	25.22	30.10	32.20	34.52	36.12
Curvlet-LBP [11]	25.66	30.28	31.49	35.49	38.26
Contourlet-LBP [12]	27.22	31.33	32.49	36.55	40.60
SRC-GSLBP [14]	31.24	33.37	34.19	38.20	42.05
GTCLBPSRC [15]	31.25	32.55	34.42	38.63	42.30
ADWTLBP [17]	33.90	39.40	41.09	42.00	45.60
DIWTLBP [18]	34.65	40.36	41.78	42.17	45.81
IDW-LBP (Proposed Method)	35.89	41.39	42.20	43.31	46.12

Table 1.
Rank-One Recognition Results of different methods on the CASIA-WebFace face database (%).

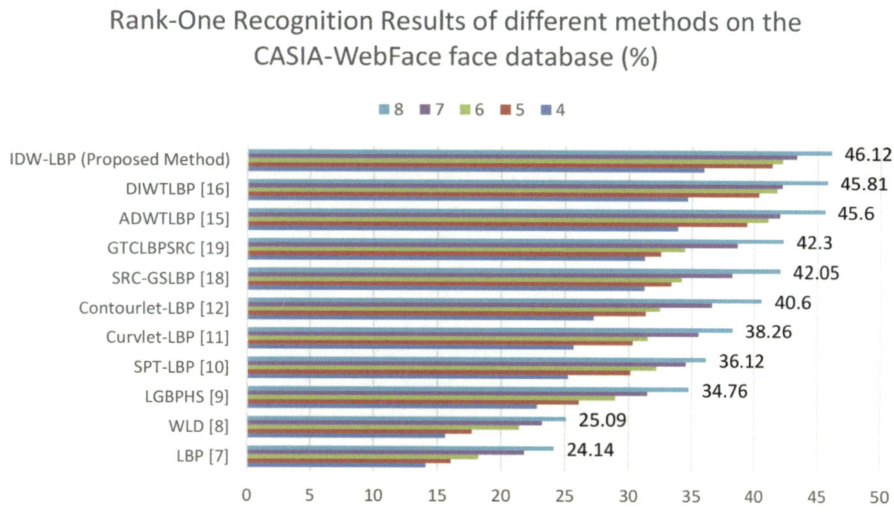

Figure 1.
Rank-One Recognition Results of different methods for the CASIA-WebFace face database (%).

illumination, expression, pose, occlusion, and age difference, this database imposes an immense challenge. **Figure 1** depicts the trend of the rank one recognition rates for different comparative methods along with the proposed method for Casia-WebFace face database.

5.2 Experiments on the labelled faces in the wild (LFW) face database

The LFW [20] is a large database that contains face images of 5749 famous personalities captured in an unrestrained environment with an extreme variation of background, pose, illumination, expression, and accessories. This makes it a challenging database for face identification. Here, we used the LFW-a database [21] which is an aligned version of the LFW database. For our experimentation purpose, we created a subset with 15 dissimilar images of 150 subjects from the LFW-a database. Each image is resized to 128×128 pixels.

Method	Number of training samples per subject				
	4	5	6	7	8
LBP [7]	24.07	30.16	38.22	42.81	44.14
WLD [8]	25.12	31.42	40.10	43.60	46.12
LGBPHS [9]	33.76	35.15	42.05	45.34	49.06
SPT-LBP [10]	34.90	40.20	43.62	45.93	50.61
Curvlet-LBP [11]	33.11	39.48	43.55	44.60	50.08
Contourlet-LBP [12]	35.09	40.25	45.52	46.41	53.32
SRC-GSLBP [14]	38.30	41.40	47.22	55.00	57.69
GTCLBPSRC [15]	40.72	42.80	50.30	58.59	60.23
ADWTLBP [17]	41.60	44.50	52.28	59.44	62.74
DIWTLBP [18]	40.80	44.89	53.03	60.18	63.04
IDW-LBP (Proposed Method)	40.77	45.64	54.80	61.01	64.42

Table 2.
Rank-One Recognition Results of different methods on the LFW face database (%).

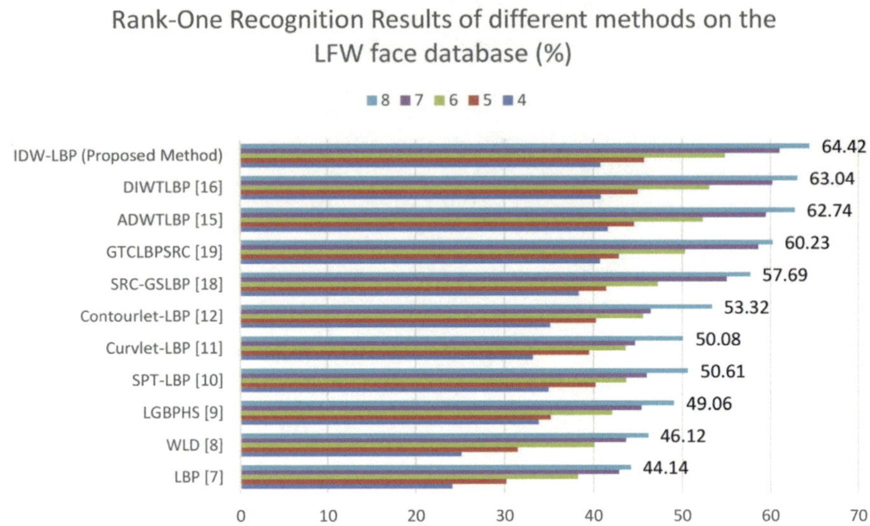

Figure 2.
Rank-One Recognition Results of different methods for the LFW face database (%).

A subset with T (T = 4, 5, 6, 7, 8) images per subject is randomly selected to form a training set, and rest images per subject are selected to form the testing set. Rank-one recognition results of different comparative methods are tabulated in **Table 2**. Since the images are selected in the unrestrained environment the Rank-one recognition results are also low in this database.

Figure 2 depicts the trend of the rank one recognition rates for different comparative methods along with the proposed method for LFW face database.

6. Conclusion

This chapter discusses the recently developed implementation of interpolation-based ADWT with seven directions and an improved QTP scheme to extract directional MRA features from face images. LBP is applied to the selected top-level IDW sub-bands to extract the multi-region histogram-based local descriptive features. Experiments conducted on the complex face databases such as CASIA-WebFace and LFW database exhibit the efficacy of our proposed method. The identification results of our method are compared with various methods which include local descriptors such as LBP and WLD. Few LBP-based non-adaptive MRA methods are also utilized for a fair comparison as our method also falls into the category of MRA based methods.

LBP and WLD suffer from issues such as the large size of histogram features, extraction of only very local texture details, limited discriminative ability, and intolerance to noise.

It also is evident from **Tables 1** and **2** that LGBPHS, SPT-LBP, Curvelet-LBP, and Contour-LBP methods provide lesser results against our proposed method. We examine progress over SRC-GSLBP, GTCLBPSRC which illustrates some effectiveness of usage of sparse features at the cost of increased complexity in implementation.

SPT-LBP also exhibits comparable performance to the proposed method but the feature selection is threshold dependent and necessitates the selection of sub-bands for efficient feature extraction. Curvelet-LBP uses only the LBP coded image of the approximation sub-band and mid-frequency sub-bands coefficients for feature generation and does not consider the multi-region information. Thus LGBPHS, Curvelet-LBP, Contourlet-LBP, and SPT-LBP are both memory and time exhaustive to extract the multiresolution and multi-orientation features due to selection and feature extraction from different sub-bands.

Moreover, these methods despite capturing the directional information lack the adaptation in selecting the directional details based on the image description and suffer from various issues such as the selection of sub-bands, high computational rate, and complex filter design. The GTCLBPSRC delivers close results to the proposed method for all the databases but at cost of additional computational time and due to the implementation of Gabor wavelet transform (GWT) which exhibits over-complete representation. We also examined comparable progress over SRC-GSLBP and GTCLBPSRC for all the databases which illustrate the effectiveness of sparse representation methods. The IDW method consists of benefits such as directional lifting and adaptation in the direction selection as per the characteristics of the images within a block of samples. Moreover, as a result of lifting based factorization, perfect reconstruction is also assured and the resultant multiresolution image is completely compatible with that of the conventional 2-D DWT multiresolution image. These facts effectively consider various edge manifolds that represent different face variations.

We also compared the proposed method with recently developed methods which also considers the facial descriptions in an adaptive MRA-based structure such as ADWT [17] and DIWT [18]. We applied a similar procedure to extract the LBP based features. The improvement in our method is visible owing to the adaptation of more directions as compared to DIWT [18] and application of sub-pixel interpolation in IDW which is absent in the ADWT method.

Thus, as per **Tables 1** and **2**, it is verified that the IDW method exhibit high discrimination capability and offers excellent recognition results for a very complex database such as CASIA-WebFace and LFW databases which consists of facial variations with mild to intense pose, expression, and illumination variations. Experiments performed for an identification process verify that the proposed method excels with all the comparative methods.

Author details

Mohd. Abdul Muqeet[1]* and Qazi Mateenuddin Hameeduddin[2]

1 Faculty of Electrical Engineering Department, Muffakham Jah College of Engineering and Technology, Hyderabad, Telangana, India

2 Faculty of Electronics and Communication, India Naval Academy, Ezhimala, Kannur, Kerala, India

*Address all correspondence to: ab.muqeet2013@gmail.com

References

[1] Anil K. Jain, Arun Ross, and Sharath Pankanti, Biometrics: A Tool for Information Security, IEEE Transactions on Information Forensics and Security, Vol. 1, no. 2, June 2006.

[2] M. Turk and A. Pentland, Eigenfaces for recognition, J. Cognitive Neuroscience., vol.3, no.1, pp.71–86, 1991.

[3] P. N. Belhumeur, J. P. Hespanha, and D. J. Kriegman, Eigenfaces versus Fisherfaces: Recognition Using Class Specific Linear Projection, IEEE Trans. on Pattern Analysis and Machine Intelligence, vol. 19, no. 7, pp. 711–720, Jul. 1997.

[4] Mallat S. A theory for multiresolution signal decomposition: The wavelet representation, IEEE Transactions on Pattern Analysis and Machine Intelligence, 11(7): 674–693, July 1989.

[5] M. A. Muqeet and R. S. Holambe, Face identification using LDA based generalized half band polynomial wavelet filter bank, 2016 International Conference on Electrical, Electronics, and Optimization Techniques (ICEEOT), Chennai, India, 2016, pp. 4649-4653. DOI: 10.1109/ICEEOT.2016.7755601.

[6] Mohd. Abdul Muqeet, Raghunath S. Holambe, Enhancing Face identification Performance using Triplet Half Band Wavelet Filter Bank, International Journal of Image, Graphics, and Signal Processing (IJIGSP), vol.8, no.12, pp.62-70, 2016. DOI: 10.5815/ijigsp.2016.12.08.

[7] T. Ojala, M. Pietikainen, T. Maenpaa, Multiresolution gray-scale, and rotation invariant texture classification with local binary patterns, IEEE Trans. Pattern Anal. Mach. Intell. 24 (7) (2002) 971–987.

[8] J. Chen, S. Shan, C. He, et al., WLD: a robust local image descriptor, IEEE Trans. Pattern Anal. Mach. Intell., vol. 32, no. 9, pp. 1705-1720, 2010.

[9] W. Zhang, S. Shan, W. Gao, H. Zhang, Local Gabor binary pattern histogram sequence (LGBPHS): a novel non-statistical model for face representation and recognition, in Proceedings of IEEE International Conference and Computer Vision, 2005, pp. 786–791.

[10] A. Alelaiwi et al., Steerable pyramid transform, and local binary pattern based robust face identification for e-health secured login, Computers and Electrical Engineering (2016), http://dx.doi.org/10.1016/j.compeleceng.2016.01.008.

[11] L. Zhou, W. Liu, ZM. Lu, T. Nie, Face identification based on curvelets and local binary pattern features via using local property preservation, Journal of Systems and Software. 95: 209-216. DOI: 10.1016/j.jss.2014.04.037.

[12] H. Y. Patil, A. G. Kothari, K. M. Bhurchandi, Expression Invariant Face identification using Local Binary Patterns and Contourlet Transform, Optik, vol. 127, pp. 2670-2678, 2016.

[13] Y.Z. Goh, A.B.J. Teoh, M.K.O. Goh, Wavelet local binary patterns fusion as illuminated facial image preprocessing for face verification, Expert. Syst. Appl., vol. 38, pp. 3959-972, 2011.

[14] Z. Liu, X. Song, Z. Tang, A novel SRC fusion method using hierarchical multi-scale LBP and greedy search strategy, Neurocomputing, vol. 151, pp.1455-1467, 2015.

[15] X. Wang, Q. Zhu, J. Cui, Y. Wang, Sparse representation method based on Gabor and CLBP, Optik, vol. 124, pp. 5843-5850, 2013.

[16] Maleki, A., Rajaei, B, Pourreza, H. R., Rate-Distortion Analysis of Directional Wavelets, Image Processing, IEEE Transaction on Image Processing, vol.21, no.2, pp.588-600, Feb. 2012.

[17] M. A. Muqeet, R. S. Holambe, Local appearance-based face identification using adaptive directional wavelet transform, J. King Saud Univ.-Computer Inform. Sci., vol. 31, pp. 161-174. (2019).

[18] Muqeet, Mohd. Abdul, Holambe, R. S., Local binary patterns based on directional wavelet transform for expression and pose-invariant face identification, Applied Computing and Informatics, vol.15, Issue. 2, July 2019, Pages 163-171.

[19] M. A. Muqeet, R. S. Holambe, A collaborative representation face classification on separable adaptive directional wavelet transform based completed local binary pattern features, Eng. Science and Tech., an Intern. Journ., vol. 21, no. 4, pp. 611-624.

[20] G.B. Huang, M. Ramesh, T. Berg., E. Learned-Miller, Labeled face in the wild: a database for studying face recognition in unconstrained environments, Technical Report, 07-49, Univ. of Massachusetts, Amherst, 2007.

[21] Wolf, L., Hassner, T., Taigman, Y., Similarity scores based on background samples, Computer Vision- ACCV 2009, pp. 88-97, 2010.

[22] D. Yi, Z. Lei, S. Liao, S.Z. Li, Learning face representation from scratch, arXiv preprint arXiv:1411.7923, 2014.

Chapter 4

Region of Interest Localization Methods for Publicly Available Palmprint Databases

Xu Liang, Dandan Fan, Zhaoqun Li and David Zhang

Abstract

So far, there exist many publicly available palmprint databases. However, not all of them have provided the corresponding region of interest (ROI) images. If everyone uses their own extracted ROI images for performance testing, the final accuracy is not strictly comparable. Since ROI localization is the critical stage of palmprint recognition. The location precision has a significant impact on the final recognition accuracy, especially in unconstrained scenarios. This problem has limited the applications of palmprint recognition. However, many currently published surveys only focus on feature extraction and classification methods. Throughout these years, many new ROI localization methods have been proposed. In this chapter, we will group the existing ROI localization methods into different categories, analyze their basic ideas, reproduce some of the codes, make comparisons of their performances, and provide further directions. We hope this could be a useful reference for further research.

Keywords: biometrics, palmprint recognition, palmprint database, region of interest localization, palm region segmentation

1. Introduction

Palm-related biometrics can easily reach high accuracy due to two reasons. One is that palmprint contains plenty of features, such as principal lines, wrinkles, ridges and valleys, and minutiae points; another one is that the regions of interest (ROIs) could be aligned with the help of the finger valley points. Since the captured palms may have different rotations and scales, to obtain high accuracy, the extracted palmprint images should be aligned with each other. It means the palmprint region should be localized based on the relative coordinate system, which is established basing on the keypoints of the finger valleys. Most of the current palmprint recognition algorithms are based on the direction information of the palmprint lines and textures [1, 2]. Hence, misalignment will significantly affect the final matching score. A robust and precise ROI localization method is essential for palmprint recognition, especially for touchless applications. Many organizations have collected their palmprint databases based on different research targets. More and more novel databases arise in recent years; some of them are captured across different devices, some with different illuminations, and some at different distances.

In the following section, we will review the current palmprint databases and ROI localization methods.

2. Palmprint databases and ROI localization methods

2.1 Comparisons of the current palmprint databases

Table 1 summarizes the current palmprint databases. Some basic information is compared. In **Table 1**, official ROI means whether the official ROI images are provided; localization code means whether the corresponding ROI extraction code is released. Some sample images of these databases are shown in **Figure 1**.

2.1.1 The Hong Kong Polytechnic University (PolyU) Palmprint Database

The PolyU Palmprint Database is the first publicly available palmprint database. It contains 7752 images captured from 386 hands in two sessions; around 10 images are collected for each palm in each session. The palmprint acquisition device is a contact-based device that consists of a high-quality industrial monochrome camera and a well-selected ring light source. The palm pose also is restricted by the pegs. So the captured images have high image qualities. However, the released image resolution, 384 × 284 pixels, is relatively low.

2.1.2 PolyU Palmprint Multi-spectral Database

Authentication by just RGB or gray images may not be safe enough; attacks from fake palm images and videos can easily spoof the system. Hence, the multi-spectral based palmprint recognition starts to draw attention. Four spectrums, red, green, blue, and near-infrared (NIR), are utilized to establish the PolyU Multi-spectral Palmprint Database. A contact-based device is employed to capture images from 250 volunteers in two sessions. 24,000 images are captured from 500 palms. Each palm contributes six images in each session under each spectrum. Our observation shows that the images captured under blue light have the highest sharpness, while that captured under NIR light has the lowest.

2.1.3 CASIA Palmprint Image Database

In CASIA Palmprint Database, 5502 images are captured from 312 subjects, around 9 images for each palm. The authors also made their own image acquisition device, which has a big enclosure and a black backboard. During the capture process, the user puts his/her palm into the enclosure back on the unicolor board; the ambient light is blocked by the enclosure. Hence, the palm images are captured in an ideal environment, but the sharpness of the palm images is not very high. Besides, some palm images are captured with significant rotations, and some fingers have moved out of the imaging window. These factors make it difficult to localize the palmprint ROIs.

2.1.4 CASIA Multi-spectral Palmprint Image Database V1.0

CASIA-MS-Palmprint V1 is a touchless multi-spectral palmprint database collected under six spectrums in two sessions. The 460, 630, 700, 850, 940 nm and WHITE spectrums are employed in their self-developed device. However, the image sharpness also is not very high compared with the PolyU multi-spectral

Dataset	No. of palms	No. of images	No. of sessions	Image size	Format	Official ROI	Localization code	Ref.
PolyU	386	7752	2	384×284	BMP	×	×	[3, 4]
PolyU-MS	500	24,000	2	352×288	JPG	✓	×	[5, 6]
CASIA	624	5502	1	640×480	JPG	×	×	[7, 8]
CASIA-MS	200	7200	2	768×576	JPG	×	×	[9, 10]
IITD	470	2601	1	1600×1200	BMP	✓	×	[11–13]
PolyU-IITD	1400	14,000	—	624×468	JPG	—	×	[14]
COEP	167	1305	1	1600×1200	JPG	×	×	[15]
KTU	145	1752	1	768×576	BMP	✓	×	[16, 17]
GPDS	100	2000	1	1200×1600	BMP	✓	×	[18, 19]
Tongji	600	12,000	2	800×600	TIFF	✓	×	[20–22]
MPD	400	16,000	2	$3120 \times 4160, 3120 \times 4208$	JPG	✓	✓	[23]
NTU-CP-v1	655	2478	2	$420 \rightarrow 1977$	JPG	×	×	[24]
NTU-PI-v1	2035	7781	—	$30 \rightarrow 1415$	JPG	×	✓	[24, 25]

Table 1.
Comparisons of the publicly available palmprint datasets.

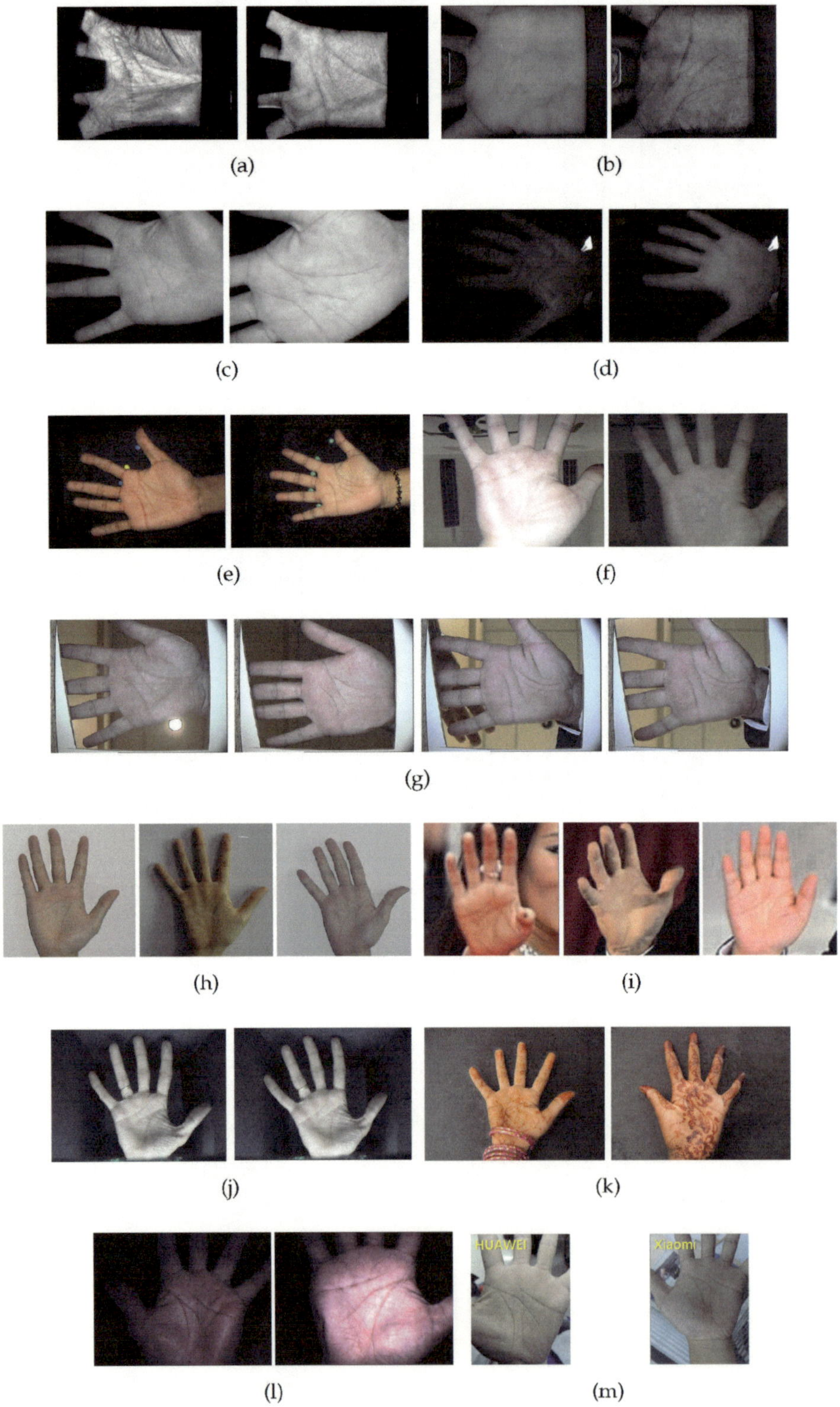

Figure 1.

Image samples of different databases. (a) PolyU Palmprint Database; (b) PolyU Multi-spectral Palmprint Database; (c) CASIA Palmprint Database; (d) CASIA Multi-spectral Palmprint Database; (e) COEP Palmprint Database; (f) GPDS100Contactlesshands2Band Palmprint Database; (g) KTU Touchless Palmprint Database; (h) NTU Contactless Palmprint Database; (i) NTU Palmprint Database from Internet; (j) IITD Touchless Palmprint Database; (k) PolyU-IITD Contactless Palmprint Database; (l) Tongji Palmprint Database; (m) Tongji MPD.

database in which the palm is captured by a contact-based device. This database contains 7200 palm images captured from 200 different palms. For each session, under each light spectrum, three images are captured from each palm.

2.1.5 IIT Delhi (IITD) Touchless Palmprint Database version 1.0

The palm images in the IITD palmprint database are captured with large rotation variations. The touchless imaging setup consists of a big black box, a digital camera, and a circular fluorescent light source. It provides 2601 palm images collected from 470 hands, including 1301 left-palm images and 1300 right-palm images. The official ROI images have been normalized and thus show obvious principal lines and wrinkles.

2.1.6 PolyU-IITD Contactless Palmprint Images Database version 3.0

This database is collected from the volunteers in different countries, China and India, by a general-purpose handheld camera over the years. Totally 14,000 images are captured from 1400 palms. The characteristic of this database is that the images are collected across different locations, times, occupations, and age ranges. Both normal and abnormal hands are involved (as is shown in **Figure 1(k)**).

2.1.7 COEP Palmprint Database

The palm images in this database have high resolutions. They are captured by Canon PowerShot SX120 IS, the image resolution is 1600 × 1200 pixels. According to the file's attribute information, the image reaches 180 dots per inch (DPI). Most of the ridge and valley lines could be seen clearly. During the image capture process, the palm position is restricted by five pegs, so the captured images are with low rotation degrees. It is a good dataset for studying the palmprint image sharpness. The downloaded dataset contains 1305 images pertaining to 167 palms, around 8 images per palm.

2.1.8 GPDS100Contactlesshands2Band Database

Both visible light camera and infrared (IR) light camera are adopted to collect 1000 visible light palm images and 1000 IR light palm images from 100 volunteers. Each palm contributes 10 visible light images and 10 IR light images. The user places his/her palm over the camera and touchlessly adjusts the position and pose of the hand in order to overlap with the hand mask drawn on the device screen. The image sharpness is not very high. However, it is a meaningful database because the images' qualities are more close to that captured in real-world applications.

2.1.9 KTU CVPR Lab. Contactless Palmprint Database

The author made a new device by a low-cost camera to capture palm images with a resolution of 768 × 576 at 75 DPI. Totally 1752 images are collected from 145 palms, about 12 images for each palm. The images are captured under different ambient light intensities, backgrounds, finger postures (finger space and finger rings), and different hand distances, rotations, and translations. The image sharpness level of this database is not very high.

2.1.10 Tongji Palmprint Database

It was the biggest touchless palmprint database in 2017. The authors also made a novel palmprint acquisition equipment that consists of a digital camera, a ring light source, a screen, and a vertical enclosure. This device can capture both visible light palmprint images and infrared light palm vein images. During collection, the user's palm is put into the enclosure to avoid ambient light. At the same time, the upper screen will show the palm in real time, so that the user knows how to put his/her hand and when to stop and hold. Totally 12,000 images are captured from 600 hands in 2 sessions. For each palm, in each session, 10 palmprint images are collected.

2.1.11 Tongji Mobile Palmprint Dataset

The device used in Tongji Palmprint Database provides a stable environment for palmprint acquisition; this strategy can ensure the final recognition performance. However, the big enclosure also has limited its applications. So the Tongji group further collected another novel database by the widely used mobile phones. The palm images are captured in the natural indoor environment. Two mobile phones are used, including HUAWEI and Xiaomi. This dataset contains 16,000 palmprint images from 400 palms collected in 2 sessions. In each session, each mobile phone captures 10 images for each palm. All palm images are labeled, and corresponding codes are released on the author's homepage [23].

2.1.12 Xi'an Jiao Tong University (XJTU) Unconstrained Palmprint Database

The XTJU-UP databases [26] are collected by five mobile phones, including iPhone 6S, HUAWEI Mate 8, LG G4, Samsung Galaxy Note5, and Xiaomi MI8, with and without the built-in flashlight source. There are 100 volunteers; each palm provides 10 images for each phone, under each illumination condition. So, totally 20,000 images are captured ($100 \times 2 \times 10 \times 5 \times 2$).

2.1.13 Nanyang Technological University (NTU) Palmprint databases version 1

The NTU palmprints from the Internet (NTU-PI-v1) database consists of 7781 hand images collected from the Internet. Hence, the palm images are captured in an uncontrolled and uncooperative environment. The images in it are collected from 2035 different palms of 1093 subjects with different ethnicity, sex, and age. Around four images are collected for each palm. It is the first large database established for studying palmprint recognition in the wild. But the image sharpness is relatively low compared with the normal palmprint images. The NTU Contactless Palmprint Database (NTU-CP-v1) contains 2478 palm images captured from 655 palms of 328 subjects using cameras of Canon EOS 500D or NIKON D70s, around four images for each palm. Currently, the samples for each category are relatively few compared with the other databases.

2.2 Related work on ROI localization

The most widely used ROI localization method is proposed in [4]. Its main idea is first detecting the keypoints of the finger valleys and then establishing a local coordinate system based on the detected keypoints, so that the ROI coordinates are determined based on the palm direction and position (as is shown in **Figure 2**). Most of the current ROI localization methods [10, 27–32] are based on this strategy.

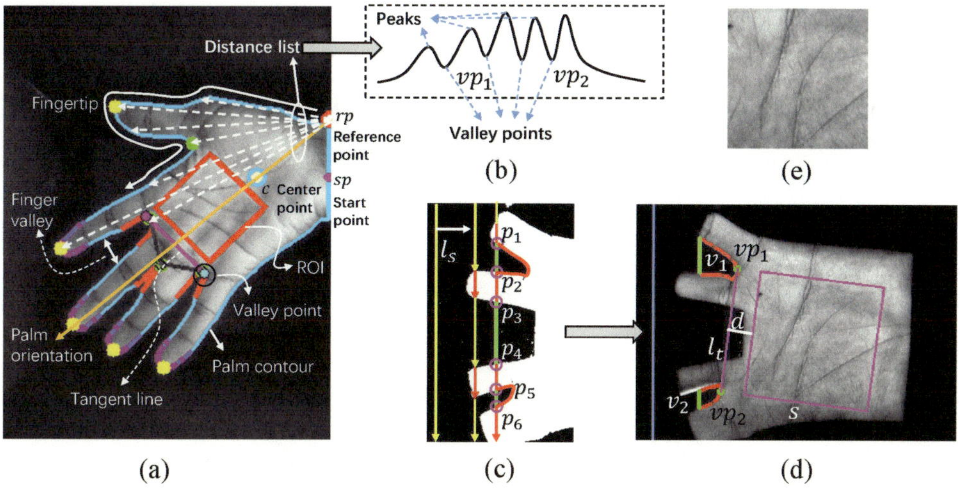

Figure 2.
The classical keypoint detection methods for ROI localization. (a) Local-extremum-based keypoint detection for a palm sample with big rotation; (b) distance curve for fingertips and finger valley points detection; (c) line-scan-based keypoint detection method; (d) notations for ROI coordinates computation; (e) the normalized ROI image.

The main problem of ROI localization is keypoint detection. There are two approaches to localize the landmarks: one is first segmenting the palm region and then searching for landmarks using the digital image processing techniques based on the detected edges; another is directly regressing the landmarks by utilizing both the hand shape and texture information.

2.2.1 Classical methods

One important goal of the first strategy mentioned above is simplifying the background. There exist three approaches:

1. Capture the palm with a unicolor backboard [4, 6, 8, 12, 21, 33]

2. Employ an IR camera or a depth camera to capture an IR image or a depth image to assist palm segmentation [34, 10, 30, 35]

3. Enhance the contrast of the foreground and background by setting a strong light source intensity and a short exposure time

Their target is enhancing the contrast of the palm region and the background. For example, in [28], the mobile phone's built-in LED flash is utilized for palm segmentation. When the flash is turned on, the palm surface is much brighter than the background, because the palm is much closer to the camera than the background. The built-in auto-exposure control function of the image signal processor (ISP) on the camera chip will automatically decrease the exposure time to capture proper palm images; the palm region should fall into the proper grayscale range. As a result, the captured background is very dark.

After hardware and acquisition mode optimization, the palm region could be segmented by skin-color thresholding or the Otsu-based methods [30, 31, 36, 37]. Maximum-connected-region detection is useful to delete the background noise. After palm region image is obtained, there are four approaches to detect the valley points:

1. Competitive valley detection algorithm [35], which traverses each contour pixel by testing and comparing its neighbor pixels' grayscale values. After palm segmentation, a binary palm image is obtained. The pixel on the palm contour is tested, taking the current contour pixel as the center point, and then 4, 8, and 16 testing points are placed around it, respectively. If in all the three tests, the pixels' values meet the predefined conditions, a line will be drawn from the center point toward the non-hand region. If this line does not cross any hand region, this center pixel is considered as a valley location. In the same way, we can find the other candidate valley points.

2. Line-scan-based methods [4, 27, 33]. After rotation normalization, the pixels are tested through a row or a column according to the specific hand orientation. In the segmented hand image, the hand region pixels are set as white, and the background pixels are set as zero. So once the pixel value changes from white to black or from black to white, the keypoints of the finger contour are detected. Then, the finger valleys can be obtained by edge tracking (as is shown in **Figure 2(c)**).

3. Local-extremum-based methods [30, 33, 38–41]. As is shown in **Figure 2(a)**, by selecting a point as the start point, we can calculate the distances between the start point and all the palm contour points to generate a distance curve. Then, on the distance curve, the local maximum points correspond to fingertips; the local minimum points correspond to finger valley points. The finger valleys could be segmented from the palm contour around the detected valley points. Then the tangent line of the two finger valleys can be detected as the reference line.

4. Convex hull-based methods [28, 42, 43]. The minimum polygon is detected to encapsulate the palm contour. Generally, the fingertips are vertexes of the convex hull. Then, the finger valleys and the valley points could be obtained as the methods mentioned above.

Generally, after the four finger valley points are obtained, we should get to know whether this hand is left or right so that the two desired valley points can be determined. As to how to identify the left and right hand, literature [35] uses geometric rules of the coordinates; literature [30] utilizes the valley areas, generally, the valley area between thumbs and index finger is bigger than that between the little and ring finger; literature [41] trained a CNN to classify it; the method proposed in [44] does not need to know the left or right information.

Rotation normalization and scale normalization are two key problems lie in palmprint preprocessing. In [33], the authors analyzed the existing methods and provided their optimized solutions in palm width detection and center point generation. Rotation normalization aims to rotate all the palms to a standard direction. To determine the main direction of the palm, many methods have been proposed. In [40], principal component analysis (PCA) is utilized to estimate the rotation angle of the palm. In [17, 44], the author utilized the training set to learn a regression model which can map the landmarks' coordinates to the palm's main direction. Hence, after landmark detection, the palm direction can be obtained by the regression model. In [30], the line crossing the middle fingertip and the palm center point is treated as the palm's center line; the palm's orientation is estimated by the line's slope.

The center point of the palm could be determined by different methods, such as the centroid of the palm region [40], the center point of the palm's maximum inscribed circle [30], the point which reaches the maximum distance value after distance transform [45, 46], or the shift from the middle point of the palm width line detected based on heart line [33].

With the information of the hand rotation angle, the palm image could be normalized to the standard direction. Then, what we need to do is scale normalization, which means to determine the side length of the ROI. The work reported in [10, 17, 21, 30, 33, 47] utilized the palm width to determine the size of the ROI, while the work reported in [4, 27–29] utilized the length of the tangent line to determine the size of the ROI. In [30], the author found that big ROI performs better. Perhaps big regions can decrease the influences of the misalignment. Here, we provide two examples for better understanding the whole process of ROI extraction.

In [27], the center block (13×13 pixels) of the image is utilized to train the skin-color model, and then the palm region is segmented based on skin-color thresholding. The candidate landmarks are obtained using the method proposed in [35]. The author proposed a two-stage strategy to achieve high robustness, i.e., palms with very big rotations or imperfect hand segmentations. In the first stage, the coarse palm direction is detected. Each candidate valley point will generate its own direction angle, and the angles will be partitioned into four coarse directions, namely, up, down, left, and right. The coarse main direction is the one which has the most supporters, and the inconsistent angles will be deleted. Then, the palm direction is calculated from the remaining angles. In the second stage, the palm image is rotated so that the four fingers point to the standard direction, and then the line-scan-based method is used to track the finger valleys. Similarly, after the valley points are detected, the ROI is derived in accordance with the reference line generated by the two valley points.

In [30], the palm is segmented from the IR light palm image by the Otsu and maximum connected domain algorithms. Then the center point of the palm is determined by the maximum inscribed circle. Right of the center point, a start point could be set. Then, the two-phase keypoint detection method is utilized for detecting the finger tips and valleys. First, the distance curve is generated by the start point and the palm contour points, and the fingertip of the middle finger is then obtained. Based on the fingertip and the center point, a new reference point could be generated to replace the start point used in the first phase. Then, with the palm orientation information, a new distance curve is generated. The precise fingertips and valley points are finally detected by the extremum points of this new distance curve. The tangent line of the valleys around the two detected valley points are obtained (as is shown in **Figure 2(a)**); we scan the palm region using lines which are parallel with the tangent line. Each line provides a palm width value, and the final palm width is determined by their median value. Last, the ROI is derived according to the reference line and the palm width.

2.2.2 New-generation methods

The methods mentioned above are all based on traditional digital image processing techniques. Most of them just utilized the edge information of the palm. However, it is not sufficient and it leads the algorithms being sensitive to palm postures and background objects. In recent years, many new methods have been proposed, such as the active shape model (ASM)-based methods [48, 49], the active appearance model (AAM)-based methods [17, 29, 50], the regression tree-based methods [47], and the deep learning-based methods [24, 41]. The new-generation methods utilized both the edge and texture information to learn much more robust models to regress the landmarks. The main stages of palmprint ROI localization is detecting the palm region from the whole image, regressing the landmarks, determining the palm orientation and width, establishing local coordinate system, and computing the ROI locations.

In [17, 44], 25 hand landmarks are selected to form a shape, including 10 end points and 15 landmarks of the finger valleys and palm boundary. This shape convers the finger roots and the interdigital regions of the palm. By AAM algorithm, both the hand shape and the palm texture information are utilized, the shape and corresponding landmark points can automatically reshape itself to fit the real hand contour. To evaluate the localization performance, the authors proposed a modified point-to-curve distance and a margin width metric. Since the initial position of the shape model is critical to the regression performance, the fitting process is divided into two stages. At first five rotations and five scale factors are used to generate 25 initial shapes. After regression, only the shape models, which obtain the 15 optimal reconstruction errors, are passed to the second stage for fine-grained regression.

In [41], the authors proposed a CNN framework based on LeNet [51] to detect the finger valley points. The proposed network involves convolutional layers and fully connected layers; the output is a six-dimensional vector corresponding to the three valley points between fingers excluding the thumb. In their work, two neural networks are designed: one is for identifying whether the hand is left or right, and another is for landmark localization. According to their experiments, the first network can perfectly identify the hand being a left or right hand, and the landmark localization performance is better than the classical method which is based on Otsu segmentation and Zhang's ROI localization algorithm [4].

In [24], based on VGG-16 [52], the authors designed an end-to-end neural network to localize the hand landmarks, generate the aligned ROI, and do feature extraction and recognition tasks at the same time. The hand region is extracted from the original Internet image, and then it is resized to 227 × 227 pixels. The normalized palm image will be put into the designed CNN for aligned ROI localization, feature extraction, and classification. The proposed network consists of two subnetworks, ROI Localization and alignment network (ROI-LAnet) and feature extraction and recognition network (FERnet). More than three landmarks are determined by the author in order to be able to parametrize non-rigid transformations.

In [47], at first the palm position is detected by techniques of sliding window, histogram of oriented gradient (HOG) [53], and support vector machine (SVM). In the training set, 14 landmark points are determined and labeled manually. After landmark point regression, the reference line is established by the two valley points. The position of the center point and the side length of the ROI both are determined by the palm width.

However, there still exist some challenging problems in ROI localization waiting for better solutions:

1. Palm region segmentation under complex backgrounds

2. Keypoint detection on palms with closed or incomplete fingers

3. Keypoint detection on palms with big rotations

4. Left and right hand detection on palms having long thumbs

5. Palm scale (palm width) determination under various palm and finger poses in touchless scenarios

Compared with the new-generation methods which need to label the landmarks manually and train the regression model, the classical methods based on hardware and capture mode optimization and digital image processing algorithms are easier to use. They also can achieve high localization precisions due to its strict imaging conditions. Hence, in this chapter we still utilize the classical methods to extract the ROIs for different palmprint databases.

3. The method

3.1 The ROI localization method

As discussed above, palmprint localization involves four main stages: (1) palm region segmentation, (2) palm contour and finger valley landmark detection, (3) ROI coordinate computation, and (4) abnormal detection. The method used in this chapter is modified from [4, 30]. For palm segmentation, Otsu-based methods can achieve good results in IR image, but for visible light image, the segmentation results will be interfered by the shadow regions on the palm surface. To achieve high success rates of ROI localization, the skin-color based classifier is utilized to separate the palm region. The main work of landmark localization is detecting the finger valleys of the index-middle fingers and the ring-little fingers (as is shown in **Figure 2(a)** and **(d)**). For contact-based palmprint image, the palm position is restricted by the pegs, so the finger valleys can be easily localized by line-scan-based methods (as is shown in **Figure 2(c)**). The pixels are tested from up to down, once the value changes from **white** to **black**, point p_1 is detected. Keep testing, once the pixel value changes from **black** to **white**, point p_2 is detected. The other keypoints $(p_3 - p_6)$ also are searched in this way. If $p_1 - p_6$ cannot be detected in the current column, the scan line l_s will move from left to right by a predefined step to test the new column. This process will be iteratively conducted before $p_1 - p_6$ are detected. Then, finger valleys $v1$ and $v2$ can be obtained by edge tracking. vp_1 and vp_2 are the detected tangent points. This scan-line-based method is effective for contact-based palm images. However, in touchless environment, most palms are captured with obvious rotations, and the space between fingers also varies a lot. Hence, the line-scan-based methods are not always workable. To deal with palm rotations, hand direction should be detected first. Hand orientation can be represented by the direction of the principal axis of the palm region pixels after principal component analysis or by the direction of the line which passes through the palm center point c and the fingertip of the middle finger (as is shown in **Figure 2(a)**). The PCA-based method can be combined with the line-scan-based method to achieve simple detection for rotated palms. But we prefer the second strategy, because if someone's thumb is as long as the little finger, it may decrease the performance of the line-scan-based method. Based on the methods proposed in [30], after palm segmentation and contour detection, the reference point rp is determined. Connecting rp with each point on the palm contour in anti-clockwise direction, for each pair of points, we could calculate their distance, and then a distance list is obtained after all the points are traversed. The distance curve is shown in **Figure 2(b)**. The extremum points of this curve correspond to fingertips and valley points. Then the finger valleys can be obtained on the palm contour. After v_1 and v_2 are obtained, the tangent line can be detected by the method proposed in [4]. Its length is denoted as l_t. Based on the tangent line, the palm coordinate system could be established. d stands for the distance between the ROI and the tangent line; d can be determined by length l_t. The ROI side length s also can be determined according to l_t. Let $d = \alpha \cdot l_t$, and $s = \beta \cdot l_t$. Different values of α, β may result in different recognition performance.

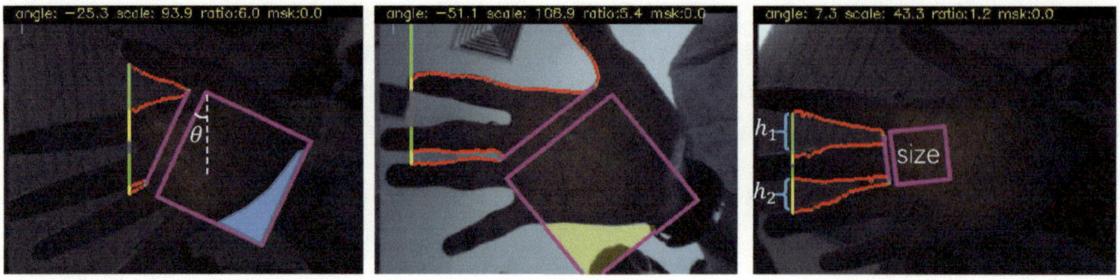

Figure 3.
Abnormal ROIs caused by complex background objects, difficult hand poses, and bad illuminations.

3.2 Abnormal detection and iterative localization

The keypoint detection method described above is based on local vision. The algorithms know whether the predefined keypoints, $p_1 - p_6$ and $vp_1 - vp_2$, are obtained. But they cannot tell whether the detected points are the correct ones. Background noise and abnormal palm poses will cause error localizations and thus generate abnormal ROIs. Those ROI images should be removed to avoid security risks. For example, during the process of sample registration, if a ROI falsely located in the background region, the black ROI image will be extracted and registered. This may cause big risk in real-world applications, since everyone can pass the system by a black image. Many deep learning-based image denoising methods have been proposed [54, 55], but, to some extent, they are too time-consuming to the target of palmprint image preprocessing. Hence, a high-speed abnormal ROI detection method is required. Here, the angle and scale of the ROI, the ratio of the background region (if there exist background regions in the located ROI), and the ratio of the width of the two finger valleys are selected as features for abnormal detection. They are denoted as θ, l_t, r_{bg}, and r_h, respectively. The area of the ROI stands for the scale information, so we use the tangent line length l_t instead. Then, for each time ROI localization, the feature vector $\{\theta, l_t, r_{bg}, r_h\}$ can be obtained. To train a SVM-based abnormal detector, first, conduct the simple localization algorithm described above on the training set to generate different kinds of ROIs (as is shown in **Figure 3**). Then, separate the ROIs into normal and abnormal subsets. Last, a binary classifier can be trained by them. According to our experiments, all the false ROIs in **Figure 3** can be successfully detected. With the abnormal detector, for line-scan-based method, once the current localized ROI is refused by the detector, it can move to the next position to iteratively detect the ROI. If the terminal condition is triggered, it means this image sample is unprocessable.

4. Experiments on different databases

In this section, the performance of the designed method is tested on different palmprint databases. More details and further updates can be found at [56]. The error rates of ROI localization are enumerated in **Table 2**. For IITD and COEP, the numbers in hard samples stand for PalmID_SampleID; for PolyU, the numbers stand for PalmID. After each experiment, the error cases are analyzed in detail.

4.1 ROI extraction for IITD Touchless Palmprint Database

Figure 4 shows the ROIs localized by the proposed method. Although the palm images in IITD Touchless Palmprint Database are captured in a black box, it is still

Database		Error rate (%)	Hard samples
IITD	Left	0.38	0037_0006, 0107_0002, 0152_0003, 0181_0001, 0209_0003
	Right	0.31	0137_0001, 0140_0004, 0204_0003, 0204_0004
COEP		0.31	0103_0004, 0145_0003, 0159_0006, 0164_0002
PolyU		0.54	0004, 0039, 0073, 0109, 0127, 0187, 0223, 0224, 0245, 0246, 0259, 0271, 0273, 0287, 0293, 0307, 0311, 0328, 0379

Table 2.
ROI localization results of different databases.

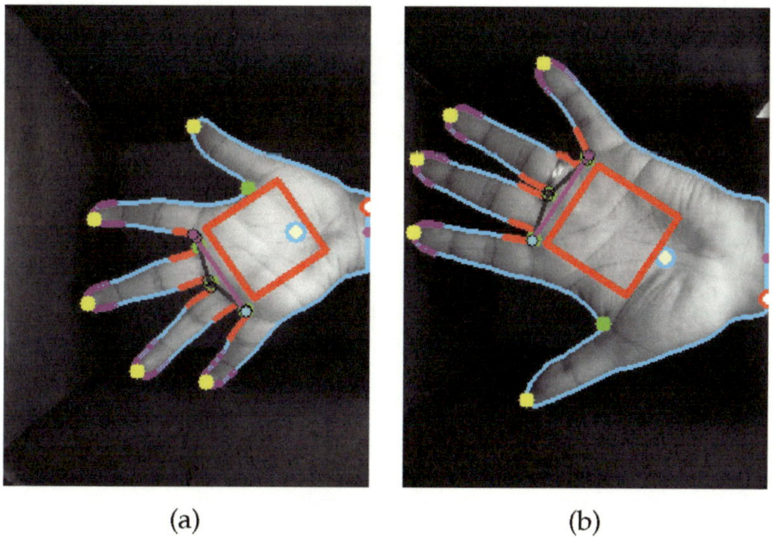

Figure 4.
ROI localization results on IITD. (a) Left hand; (b) right hand.

difficult to segment the palm region. The strong light source and the small enclosure lead to light reflections and ray occlusions, which generate many bright regions in the background and dark regions on the palm surface. Hence, the brightness information is not sufficient for segmenting the palm region. The color information should be utilized. As is shown in **Figure 5**, we randomly cropped some palm skin and background image patches to build a training set for segmentation. A SVM-based binary classifier is learned from the training dataset; the segmented palm

Figure 5.
Image patches cropped from the palm skin and the black box. (a) Patches of the palm surface; (b) patches of the black box.

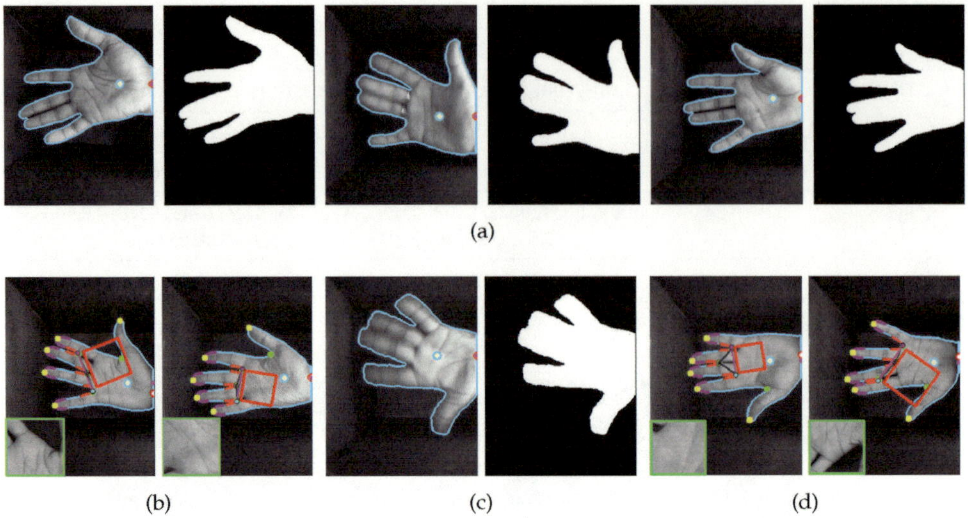

Figure 6.
Images cannot be localized in IITD database. (a) and (c) are the finger detection failed samples; (b) and (d) are the ROI localization error samples.

region can be seen in **Figure 6**. For palm skin patches, both the bright and dark regions are selected to learn a precise classification plane. In the color space, the palm can be easily segmented from the unicolor background. After palm region segmentation, vp_1 and vp_2 are detected by the local-extremum-based method. **Results: Figure 6** shows the IITD samples that are hard to process. If we cannot detect five fingertips and four finger valley points after palm segmentation, the system will directly return by giving an error code (as is shown in **Figure 6(a)** and **(c)**). If false keypoints are detected due to difficult palm poses, the finally extracted ROI images are abnormal images which should be discarded in real-world applications (as is shown in **Figure 6(b)** and **(d)**).

4.2 ROI extraction for COEP Palmprint Database

The line-scan-based keypoint detection method is used for COEP. **Figure 7** shows the ROI localization results on COEP database. Since the pegs used in their imaging setup may interfere the keypoint detection algorithm, we should delete

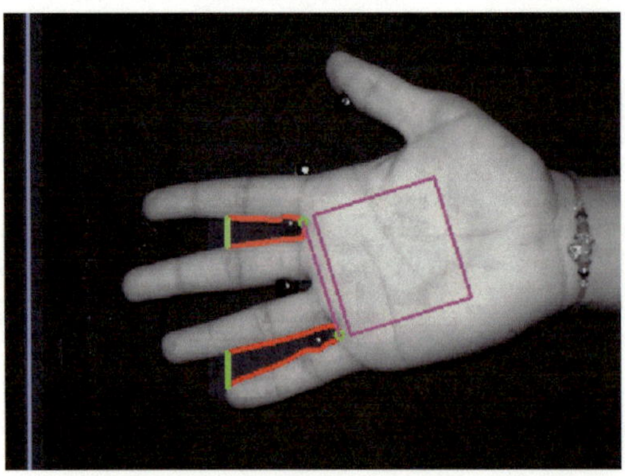

Figure 7.
ROI localization result of the COEP database.

them first. The pegs' colors are green, blue, and yellow. After removing the bright yellow pixels in the image, we extract the red channel from the original RGB image to conduct ROI localization algorithm. In this way, the green and blue pegs can be automatically removed (as is shown in **Figure 7**). **Results:** as is shown in **Figure 8**, after ROI localization, four images failed to be correctly localized. All of the four error cases are caused by closed fingers.

4.3 ROI extraction for PolyU Palmprint Database

For PolyU database, which contains 7752 images, the line-scan-based method is utilized to localize the ROI. At last, 42 samples failed to be localized. As is shown in **Figure 9**, most of them are caused by small finger valleys (palm pose) and unideal palm region segmentations (only grayscale information can be utilized). In **Table 2**, only the user ID is listed for the PolyU database.

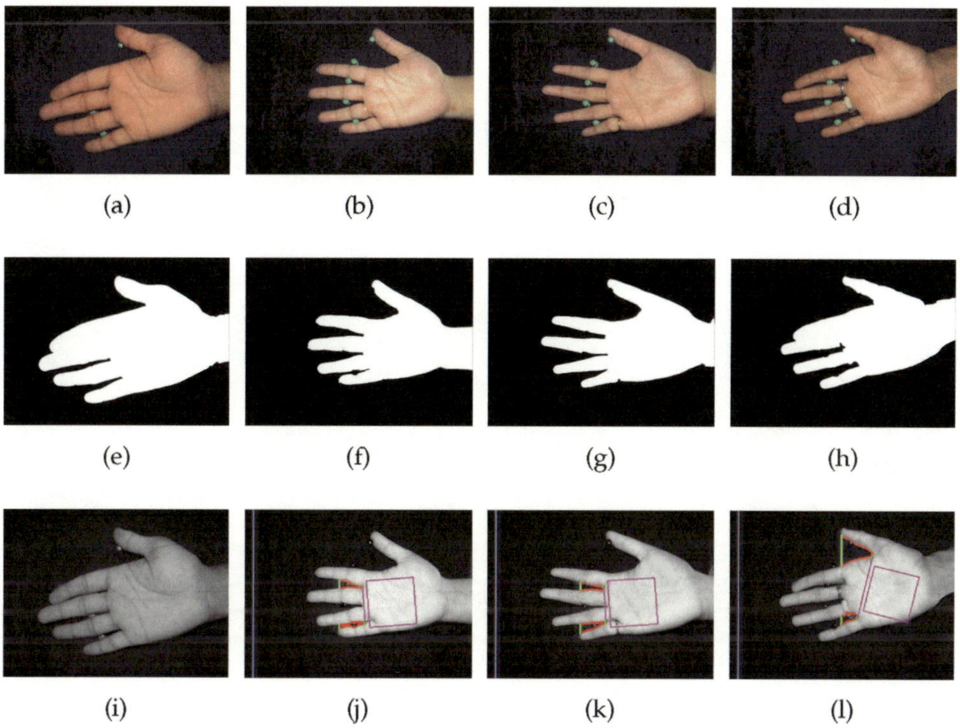

Figure 8.
Images cannot be localized of COEP database. (a)–(d), (e)–(h), and (i)–(l) are the original, binary, and ROI localization images, respectively. The image ID of (a), (e), and (i) is 0103_0004; the image ID of (b), (f), and (j) is 0145_0003; the image ID of (c), (g), and (k) is 0159_0006; and the image ID of (d), (h), and (l) is 0164_0002.

Figure 9.
Hard samples of PolyU.

5. Conclusions

The motivation of this chapter is providing a uniform ROI localization method to extract standard ROI images. This is very meaningful for comparing the new proposed feature extraction and identification algorithms. This also can lower the threshold of the palmprint research for beginners, because preprocessing is very complex and time-consuming. The method used in this chapter is not for real-world applications; it is only a ROI extraction tool for the publicly available databases. According to this goal, the simple method, based on classical digital image processing and machine learning techniques, is selected in this chapter.

Acknowledgements

We would like to thank all the volunteers who contributed their palm images to establish these palmprint databases and thank all the organizations who shared their databases. This work is supported in part by the Shenzhen Institute of Artificial Intelligence and Robotics for Society.

Author details

Xu Liang[1,3], Dandan Fan[2,3], Zhaoqun Li[2,3] and David Zhang[1,2,3,4]*

1 Harbin Institute of Technology, Shenzhen, China

2 The Chinese University of Hong Kong, Shenzhen, China

3 Shenzhen Institute of Artificial Intelligence and Robotics for Society, Shenzhen, China

4 School of Science and Engineering, The Chinese University of Hong Kong, Shenzhen, China

*Address all correspondence to: davidzhang@cuhk.edu.cn; csdzhang@comp.polyu.edu.hk

References

[1] Fei L, Lu G, Jia W, Teng S, Zhang D. Feature extraction methods for palmprint recognition: A survey and evaluation. IEEE Transactions on Systems, Man, and Cybernetics: Systems. 2019;**49**(2):346-363

[2] Jia W, Zhang B, Lu J, Zhu Y, Zhang Y, Zuo W, et al. Palmprint recognition based on complete direction representation. IEEE Transactions on Image Processing. 2017;**26**(9): 4483-4498

[3] PolyU Palmprint Database. Available from: https://www4.comp.polyu.edu.hk/biometrics/ [Accessed: 29 May 2020]

[4] Zhang D, Kong WK, You J, Wong M. Online palmprint identification. IEEE Transactions on Pattern Analysis and Machine Intelligence. 2003;**25**(9): 1041-1050

[5] PolyU Multi-Spectral Palmprint Database. Available from: https://www4.comp.polyu.edu.hk/biometrics/ [Accessed: 29 May 2020]

[6] Zhang D, Guo Z, Lu G, Zhang L, Zuo W. An online system of multi-spectral palmprint verification. IEEE Transactions on Instrumentation and Measurement. 2010;**59**(2):480-490

[7] CASIA Palmprint Database. Available from: http://biometrics.idealtest.org/dbDetailForUser.do?id=5 [Accessed: 29 May 2020]

[8] Sun Z, Tan T, Wang Y, Li SZ. Ordinal palmprint representation for personal identification [representation read representation]. In: Proceedings of the IEEE Computer Society Conference on Computer Vision and Pattern Recognition (CVPR'05); 20-26 June 2005; San Diego, CA. New York: IEEE; 2005. pp. 279-284

[9] CASIA Multi-Spectral Palmprint Database. Available from: http://biometrics.idealtest.org/dbDetailForUser.do?id=6 [Accessed: 29 May 2020]

[10] Hao Y, Sun Z, Tan T, Ren C. Multi-spectral palm image fusion for accurate contact-free palmprint recognition. In: Proceedings of the IEEE International Conference on Image Processing (ICIP'08); 12-15 October 2008; San Diego, California, USA. New York: IEEE; 2008. pp. 281-284

[11] IITD Palmprint Database. Available from: https://www4.comp.polyu.edu.hk/csajaykr/IITD/Database_Palm.htm [Accessed: 29 May 2020]

[12] Kumar A. Incorporating cohort information for reliable palmprint authentication. In: Proceedings of the Indian Conference on Computer Vision, Graphics & Image Processing (ICVGIP'08); 16-19 December 2008; Bhubneshwar, India. New York: IEEE; 2008. pp. 583-590

[13] Kumar A, Shekhar S. Personal identification using rank-level fusion. IEEE Transactions on Systems, Man, and Cybernetics: Part C. 2011;**41**(5): 743-752

[14] PolyU-IITD Palmprint Database. Available from: http://www4.comp.polyu.edu.hk/csajaykr/palmprint3.htm [Accessed: 29 May 2020]

[15] COEP Palmprint Database. Available from: https://www.coep.org.in/resources/coeppalmprintdatabase [Accessed: 03 May 2020]

[16] KTU Palmprint Database. Available from: https://ceng2.ktu.edu.tr/cvpr/contactlessPalmDB.htm [Accessed: 29 May 2020]

[17] Aykut M, Ekinci M. Developing a contactless palmprint authentication

system by introducing a novel ROI extraction method. Image and Vision Computing. 2015;**40**:65-74

[18] GPDS Palmprint Database. Available from: http://www.gpds.ulpgc.es/ [Accessed: 29 May 2020]

[19] Ferrer MA, Vargas F, Morales A. BiSpectral contactless hand based biometric system. In: CONATEL 2011. New York: IEEE; 2011. pp. 1-6

[20] Tongji Palmprint Database. Available from: https://sse.Tongji.edu.cn/linzhang/contactlesspalm/index.htm [Accessed: 29 May 2020]

[21] Zhang L, Yang A, Shen Y, Yang M. Towards contactless palmprint recognition: A novel device, a new benchmark, and a collaborative representation based identification approach. Pattern Recognition. 2017;**69**: 199-212

[22] Zhang L, Cheng Z, Shen Y, Wang D. Palmprint and palmvein recognition based on DCNN and a new large-scale contactless palmvein dataset. Symmetry. 2018;**10**(4):78

[23] MPD Palmprint Database. Available from: https://cslinzhang.github.io/MobilePalmPrint/ [Accessed: 29 May 2020]

[24] Matkowski MM, Chai T, Kong WKA. Palmprint recognition in uncontrolled and uncooperative environment. IEEE Transactions on Information Forensics and Security. 2020;**15**:1601-1615

[25] NTU-PI-v1 Palmprint Database. Available from: https://github.com/matkowski-voy/Palmprint-Recognition-in-the-Wild [Accessed: 29 May 2020]

[26] XJTU-UP Palmprint Database. Available from: http://gr.xjtu.edu.cn/web/bell/resource [Accessed: 29 May 2020]

[27] Franzgrote M, Borg C, Ries BJT, Büssemaker S. Palmprint verification on mobile phones using accelerated competitive code. In: Proceedings of 2011 International Conference on Hand-Based Biometrics (ICHB'11); 17-18 November 2011; Hong Kong, China. New York: IEEE; 2011

[28] Chai T, Wang S, Sun D. A palmprint ROI extraction method for mobile devices in complex environment. In: Proceedings of IEEE 13th International Conference on Signal Processing (ICSP'16); 6-10 November 2016; Chendu, China. New York: IEEE; 2017

[29] Bingöl Ö, Ekinci M. Stereo-based palmprint recognition in various 3D postures. Expert Systems With Applications. 2017;**78**:74-88

[30] Liang X, Zhang D, Lu G, Guo Z, Luo N. A novel multicamera system for high-speed touchless palm recognition. In: IEEE Transactions on Systems, Man, and Cybernetics: Systems. 2019. DOI: 10.1109/TSMC.2019.2898684. Available from: https://ieeexplore.ieee.org/abstract/document/8666082

[31] Genovese A, Piuri V, Plataniotis K, Scotti F. PalmNet: Gabor-PCA convolutional networks for Touchless Palmprint recognition. IEEE Transactions on Information Forensics and Security. 2019;**14**(12):3160-3174

[32] Leng L, Liu G, Li M, Khan KM, Al-Khouri MA. Logical conjunction of triple-perpendicular-directional translation residual for contactless palmprint preprocessing. In: Proceedings of International Conference on Information Technology: New Generations (ITNG'14); 7-9 April 2014; Las Vegas, NV, USA. New York: IEEE; 2014

[33] Jia W, Hu RX, Gui J, Zhao Y, Ren XM. Palmprint recognition across different devices. Sensors (Switzerland). 2012;**12**(6):7938-7964

[34] Han Y, Sun Z, Wang F, Tan T. Palmprint recognition under unconstrained scenes. In: Lecture Notes in Computer Science (ACCV'07); 18-22 November 2007; Tokyo, Japan. Berlin: Springer; 2007. pp. 1-11

[35] Michael GKO, Connie T, Jin ATB. Touch-less palmprint biometrics: Novel design and implementation. Image and Vision Computing. 2008;**26**(12): 1551-1560

[36] Otsu N. A threshold selection method from gray-level histograms. IEEE Transactions on Systems, Man, and Cybernetics. 1979;**SMC-9**(1):62-66

[37] Rother C, Kolmogorov V, Blake A. "GrabCut" interactive foreground extraction using iterated graph cuts. ACM Transactions on Graphics. 2004; **23**(3):309-314

[38] Yörük E, Konukoğlu E, Sankur B, Darbon J. Shape-based hand recognition. IEEE Transactions on Image Processing. 2006;**15**(7):1803-1815

[39] Lin C-L, Chuang TC, Fan K-C. Palmprint verification using hierarchical decomposition. Pattern Recognition. 2005;**38**(12):2639-2652

[40] Ito K, Sato T, Aoyama S, Sakai S, Yusa S, Aokl T. Palm region extraction for contactless palmprint recognition. In: International Conference on Biometrics (ICB'15); 19-22 May 2015; Phuket, Thailand. New York: IEEE; 2015. pp. 334-340

[41] Bao X, Guo Z. Extracting region of interest for palmprint by convolutional neural networks. In: Proceedings of the International Conference on Image Processing Theory, Tools and Applications (IPTA'16); 12-15 December 2016; Oulu, Finland. New York: IEEE; 2016. p. 16

[42] Moço NF. Smartphone-based palmprint recognition system. In: Proceedings of International Conference on Telecommunications (ICT'14); 4-7 May 2014; Lisbon, Portugal. New York: IEEE; 2014. pp. 457-461

[43] Sun X, Xu Q, Wang C, Dong W, Zhu Z. ROI extraction for online touchless palm vein based on concavity analysis. In: Proceedings of Youth Academic Annual Conference of Chinese Association of Automation (YAC'17); 19-21 May 2017; Hefei, China. New York: IEEE; 2017. pp. 1123-1126

[44] Aykut M, Ekinci M. AAM-based palm segmentation in unrestricted backgrounds and various postures for palmprint recognition. Pattern Recognition Letters. 2013;**34**(9):955-962

[45] Lee T, Höllerer T. Handy AR: Markerless inspection of augmented reality objects using fingertip tracking. In: Proceedings of 2007 11th IEEE International Symposium on Wearable Computers (ISWC '07); 11-13 October 2007; Boston, MA, USA. New York: IEEE; 2007. pp. 83-90

[46] Borgefors G. Distance transformations in digital images. Computer Vision, Graphics and Image Processing. 1986;**34**:344-371

[47] Shao H, Zhong D, Du X. Efficient deep palmprint recognition via distilled hashing coding. In: Proceedings of the IEEE Conference on Computer Vision and Pattern Recognition Workshops (CVPR Workshops'19); 16-20 June 2019; Long Beach, CA, USA. New York: IEEE; 2019. pp. 714-723

[48] Cootes TF, Taylor CJ, Cooper DH, Graham J. Active shape models: Their training and application. Computer Vision and Image Understanding. 1995; **61**:38-59

[49] Gao F, Cao K, Lu L, Yuan Y. Mobile palmprint segmentation based on improved active shape model. Journal of

Multimedia Information System. 2018; **5**(4):221-228

[50] Cootes TF, Edwards G, Taylor C. Active appearance models. IEEE Transactions on Pattern Analysis and Machine Intelligence. 2001;**23**(6): 681-685

[51] LeCun Y, Bottou L, Bengio Y, Haffner P. Gradient-based learning applied to document recognition. Proceedings of the IEEE. 1998;**86**(11): 2278-2324

[52] Simonyan K, Zisserman A. Very deep convolutional networks for large-scale image recognition. In: 3rd International Conference on Learning Representations (ICLR'15); San Diego, CA, USA; 7–9 May 2015; Available from: https://arxiv.org/abs/1409.1556 [Accessed: 29 May 2020]

[53] Dalal N, Triggs B. Histogram of oriented gradient for human detection. In: Proceedings of the 2005 IEEE Computer Society Conference on Computer Vision and Pattern Recognition (CVPR'05); 20-26 June 2005; San Diego, CA, USA. New York: IEEE; 2005. pp. 886-893

[54] Tian C, Xu Y, Fei L, Yan K. Deep learning for image denoising: A survey. In: International Conference on Genetic and Evolutionary Computing (ICGEC'18); 14-17 December 2018; Changzhou, China. Singapore: Springer; 2018. pp. 563-572

[55] Tian C, Xu Y, Li Z, Zuo W, Fei L, Liu H. Attention-guided CNN for image denoising. Neural Networks. 2020;**124**: 117-129

[56] Palmprint ROI Extractors for Publicly Available Databases. Available from: https://github.com/xuliangcs/PalmROI [Accessed: 29 May 2020]

Chapter 5

Image Sharpness-Based System Design for Touchless Palmprint Recognition

Xu Liang, Zhaoqun Li, Jinyang Yang and David Zhang

Abstract

Currently, many palmprint acquisition devices have been proposed, but how to design the systems are seldom studied, such as how to choose the imaging sensor, the lens, and the working distance. This chapter aims to find the relationship between image sharpness and recognition performance and then utilize this information to direct the system design. In this chapter, firstly, we introduce the development of recent palmprint acquisition systems and abstract their basic frameworks to propose the key problems needed to be solved when designing new systems. Secondly, the relationship between the palm distance in the field of view (FOV) and image pixels per inch (PPI) is studied based on the imaging model. Suggestions about how to select the imaging sensor and camera lens are provided. Thirdly, image blur and depth of focus (DOF) are taken into consideration; the recognition performances of the image layers in the Gaussian scale space are analyzed. Based on this, an image sharpness range is determined for optimal imaging. The experiment results are obtained using different algorithms on various touchless palmprint databases collected using different kinds of devices. They could be references for new system design.

Keywords: palmprint recognition, system design, image sharpness assessment, scale space, field of view, depth of focus

1. Introduction

Biometric identification has been widely applied in modern society, such as electronic payment, entrance control, and forensic identification. As a reliable solution for identity authentication, biological characteristics refer to the inherent physiological or behavioral characteristics of the human body, including the iris, pattern, retina, palmprint, fingerprint, face and also voiceprint, gait, signature, key strength, etc. In the last decade, we have witnessed the successful employment of recognition systems using fingerprint, iris, and face. With the development of image capture devices and recognition algorithms, palmprint recognition receives more and more attention recently. Palmprint image contains principal lines, wrinkles, ridges, and texture that are regarded as useful features for palmprint representation and can be captured with a low-resolution image [1]. Palmprint recognition has several advantages compared with other biometrics: (1) the line features and texture features in a palmprint are discriminative and robust, which can be easily fused with other hand features (dorsal hand vein, fingerprint, finger knuckle); (2) the pattern of palmprint is mainly controlled by genetic genes, when combined with palm vein information it can achieve high antispoof capability; (3) palmprint image acquisition is convenient and low-cost, and a relative low-resolution camera and a light source are sufficient to acquire the images; (4) the palmprint acquisition is hygienic and user friendly in the real applications. Based on

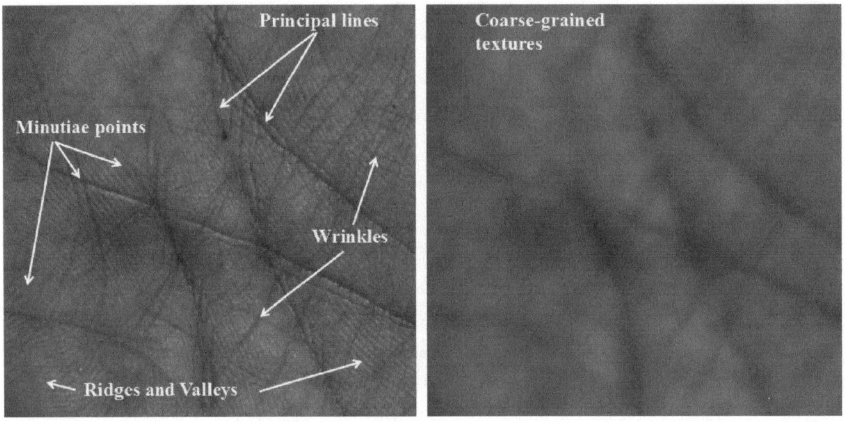

Figure 1.
Palmprint images and feature definitions.

Ref.	Year	Device type	Image type	Description
[1]	2003	Touch-based	Gray scale	Adopt low-cost camera to capture low-resolution image palmprint; use pegs as guidance
[2]	2007	Touchless	RGB and IR	Realize noncontact capturing of palmprint images under unconstrained scenes
[3]	2008	Touchless	RGB	Capture palm in real-time video stream using skin-color thresholding
[4]	2009	Touch-based	3D	Acquire depth information in palm using structured light imaging
[5]	2010	Touch-based	Multispectral	Propose an online multispectral palmprint system
[6]	2010	Touchless	RGB and IR	Capture palmprint and palm vein images simultaneously
[7]	2011	Touch-based	Gray scale and IR	Capture palmprint, palm vein, and dorsal vein images simultaneously
[8]	2012	Portable	Gray scale	Use different portable devices to capture palmprint images
[9]	2012	Touch-based	Gray scale and 3D	Acquire 3D information and 2D texture in palm
[10]	2015	Touchless	RGB	The RGB's blue and red channels are processed separately for bimodal feature extraction
[11]	2016	Touch-based	Gray scale	Develop a line scanner to capture palmprint images
[12]	2017	Touch-based	Gray scale	Proposed a novel doorknob device to capture the knuckle images
[13]	2018	Touchless	Multispectral	Capture palmprint and palm vein images in the device; established the current biggest publicly available database

Table 1.
The palmprint recognition systems.

the custom acquisition devices, more information can be retrieved in a multispectral image or 3D palmprint image. A 2D gray scale palmprint example with feature definitions is shown in **Figure 1**. The purpose of this chapter is to review recent research on palmprint acquisition systems to trace the development of palmprint recognition-based biometric systems. In this chapter, we coarsely divide the devices into three types by acquisition mode: touch-based devices, touchless devices, and portable devices. Touch-based devices usually have pegs to constrain the hand pose and position, which can capture the details of palmprint to the most extent. The illuminating environment is also stable during capturing process. These constrains ensure the captured palmprint images to be high quality. For touchless devices, users can freely place their palms in front of the camera while the hand pose is generally required to spread out the fingers. The environment during the capturing process becomes more complicated, especially the illumination. There are also datasets composed of palmprint images captured in a relatively free fashion. These images may be collected on the Internet which we will not discuss here. Otherwise, collectors use digital cameras or phone cameras to capture palmprint image, and usually, there are no strict conditions forced on the user. In the rest of this chapter, first, we will introduce the representative palmprint acquisition devices, and then study the relationship between the palm distance, image sharpness, hardware parameters, and the final recognition performance. **Table 1** summarizes the palmprint acquisition devices.

2. The current palmprint recognition devices

2.1 Touch-based devices

Reference [1] is a pioneer work for palmprint acquisition and recognition that builds the first large-scale public palmprint dataset. The captured palmprint images are low-resolution with 75 pixels per inch (PPI), so that the whole process can be completed in 1 s, which achieves real-time palmprint identification. The palmprint capture device includes a ring light source, charge-coupled device (CCD) camera, a frame grabber, and an analog-to-digital (AD) converter. Six pegs are serving as control points that constrain the user's hands. To guarantee the image quality, during palmprint image capturing, the device environment is semiclosed, and the ring source provides uniform lighting conditions. After capturing the palmprint, the AD converter directly transmits the captured images by the CCD camera to a computer. The well-designed acquisition system can capture high-quality images, which boosts the performance of the identification algorithm. The experiment result also demonstrates that low-resolution palmprint can achieve efficient person identification. Our palms are not pure planes, and many personal characteristics lie on the palm surface. From this view, 2D palmprint recognition has some inherent drawbacks. On one hand, much 3D depth information is neglected in 2D imaging. The main features in 2D palmprint are line features including principal lines and wrinkles, which is not robust to the illumination variations and contamination influence. On the other hand, the 2D palmprint image is easy to be counterfeited so that the anti-forgery ability of 2D palmprint needs improvement. For capturing depth information in palmprint, [4, 14] explores a 3D palmprint acquisition system that leverages the structured light imaging technique. Compared to 2D palmprint images, several unique features, including mean curvature image, Gaussian curvature image, and surface type, are extracted in 3D images. Many studies have proposed different algorithms that encode the line features on the palm surface; however, the discriminative and antispoof capability of palm code needs to be further improved for large-scale identification. To obtain more biometric information in the palm, in [5] a multispectral palmprint acquisition system is designed, which can capture both red, green, and blue (RGB) images and near-infrared (NIR) images of one palm. It consists of a CCD camera, lens, an A/D converter, a multispectral light source, and a light controller. The monochromatic CCD is placed at the bottom of the device to capture palmprint images, and the light controller is used to control the multispectral light. In the visible spectrum, a three-mono-color LED array is used with red peaking at 660 nm, green peaking at 525 nm, and blue peaking at 470 nm. In the NIR spectrum, a NIR LED array peaking at 880 nm is used. It has been shown that light in the 700 to 1000 nm range can penetrate the human skin, whereas 880–930 nm provides a good contrast of subcutaneous veins. The system is low-cost, and the acquired palmprint images are high-quality. By fusing the information provided by multispectral palmprint images, the identification algorithm achieves higher performance on recognition accuracy and antispoof capacity.

2.2 Touchless devices

Touch-based devices can easily capture high-quality palmprint images which contribute to high performance in person identification, while their drawbacks also lie in this acquisition mode. Firstly, users may have hygienic concerns since the device cannot be cleaned immediately. Secondly, some users may feel uncomfortable with the control pegs and constrained capture environment. Thirdly, the volume of the device is usually larger than palm, which causes problems of portability and usability. As the first attempt to solve the above issues, [2] presents a real-time touchless palmprint recognition system, and the capture processes are conducted under unconstrained scenes. Two complementary metal-oxide semiconductor (CMOS) web cameras are placed in parallel, one is a near-infrared (NIR) camera, and the other is a traditional red green blue (RGB) camera. A band pass filter is fixed on the camera lens to eliminate the influence of NIR light on the palm. The two cameras work simultaneously, and the resolution of both cameras is 640×480. For further hand detection process, during the image capture, users need to open their hands and place palm regions in front of the cameras. Also, the palm plane needs to be approximately flat and orthogonal to the optical axis of cameras. Minor in-plane rotation is allowed. The distance between the hand and device should be in a fixed range (35–50 cm) to ensure the clarity of the palmprint images. In [3], a novel touchless device with a single camera is proposed. The principle of device design is similar to [2]. During the input process, the user places his/her hand in front of the camera without touching the device, and there are no strict constraints on its pose and location. The main difference is that the paddles are placed around the camera to reduce the effect of illumination changes. By these measures, the acquisition process becomes flexible and efficient. [6] presents a touchless palmprint and palm vein recognition system. The structure of the device is similar to that in [3], which mainly contains two parallel mounted cameras with visible light and IR light. The flexibility of this touchless device is further improved. Users are allowed to position their hands freely above the sensor, and they can move their hands during the acquisition process. The acquisition program will give feedback to the user that he/she is placing his/her hand correctly inside the working volume. In this way, the device can capture high-quality palmprint and palm vein images at the same time. In [7], the palmprint, palm vein, and dorsal vein images are simultaneously captured with a touchless acquisition device. In the capturing process, the users are asked to put their hands into the device with five fingers separated. The time cost is less than 1 s. The multimodal images can be fused in the algorithm to boost the identification performance.

2.3 Portable devices

With the widespread application of digital cameras and smartphones, more and more portable biometric devices appear to us. To investigate the problem of palmprint recognition across different portable devices and build the available dataset, [8] uses one digital camera and two smartphones to acquire palmprints in a free manner.

2.4 Key problems in device design

As is discussed above, the main parts of palmprint acquisition devices are cameras and light sources. So, the problems we need to consider when designing new devices are as follows:

1. The resolution of the imaging sensor
2. The focal length of the lens
3. The distance range of the palm
4. The sharpness range of the final palmprint image
5. The light source intensity
6. The signal-to-noise ratio of the palmprint image

Many previous works have studied the light sources [15–17]. Generally, the basic goal is avoiding overexposure and underexposure. Image noise increases under low illumination conditions. Although many new deep learning-based denoising techniques are proposed [18], the most effective solution for palmprint imaging is developing active light sources to provide suitable illumination conditions. In this work, we only focus on the first four problems. We developed three palm image capture devices to test the performance of different hardware frameworks (as is shown in **Figure 2**). We denote them as $device_a$, $device_b$, and $device_c$. Among them, $device_a$ and $device_b$ are touch-based devices. $device_a$ is designed to generate high-quality palmprint images. The device contains an ultra-high-definition imaging sensor (about 500 M pixels) and a distortion-free lens. The long working distance is designed to further guarantee the image quality. During the capture process, the user's palm is put on the device to avoid motion blur. $device_b$ is designed to generate high-distortion palmprint images. It contains a high-definition imaging sensor (about 120 M pixels) and an ultrawide lens. The working distance is very short (about 2 cm). $device_c$ is a touchless device; it is designed to capture high- and low-definition images in touchless scenarios. It has two cameras, one is high-definition (120 M pixels), and the other one is low-definition (30 M pixels); both of them are equipped with distortion-free lenses. We use different devices to collect palm images from the same palm; the captured images are shown in **Figure 2(d)–(e)**. We can see that the 500 M pixel camera can capture clear ridges and valleys of the palmprint, the 120 M pixel camera can capture most of the ridges and valleys, and the 30 M pixel camera only can capture the principal lines and coarse-grained skin textures. For touchless applications, the distance between the palm and the camera is not stable. Distance variations may decrease the palm image PPI and cause defocus-blur. In practice, it is very hard to guarantee the quality of the captured images. Hence, what we want to know is which level of image sharpness is sufficient for palmprint identification.

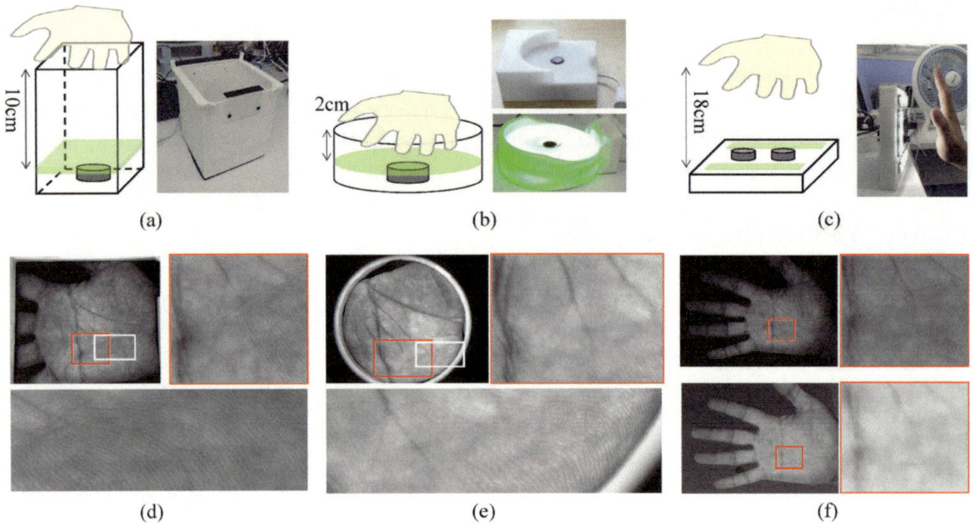

Figure 2.
Different palmprint acquisition devices and the palm images generated by them. (a) The touch-based device with a 500 M pixel imaging sensor and a long imaging distance. (b) The touch-based device with a 120 M pixel imaging sensor and a very short imaging distance. (c) The multicamera touchless device with 120 M and 30 M pixel imaging sensors and a long imaging distance. (d) The palm image captured by (a) and the corresponding enlarged local regions. (e) The palm image captured by (b) and the corresponding enlarged local regions. (f) The palm images captured by (c) and the corresponding enlarged local regions.

3. System design based on palm image sharpness

3.1 Palm distance and recognition performance

The imaging model is shown in **Figure 3**. Let l_p and w_p stand for the statistical information of the length and width of the palm, respectively. Let Z_{min} and Z_{max} stand for the minimum and maximum distance the palm can reach in the field of view (FOV). If the hand want to be captured completely, we need $l \geq l_p$ and $w \geq w_p$, where l and w are the corresponding sizes of the field of view (FOV) of the camera (as is shown in **Figure 3**). Then Z_{min} could be estimated by

$$Z_{min} = \max\left(\frac{l_p/2}{\tan\theta_u}, \frac{w_p/2}{\tan\theta_v}\right) \tag{1}$$

where θ_u and θ_v are half angles of the FOV along directions of u and v, respectively. As is shown in **Figure 3**, in the generated image, p_w (in units of pixel) is the palm width. r_w (in units of pixel) is the length of the tangent line formed by two finger valley key points. We introduce it here, because most of the region of interest (ROI) localization methods utilize those two key points [1]. The PPI is calculated by

$$ppi = p_w/w_p \tag{2}$$

in which w_p is the fixed real palm size. Based on the triangle geometry constraints defined in the pin-hole imaging model [19], we have

$$p_w/f = w_p/z \tag{3}$$

where f is the focal length (in units of pixel), which is related with the pixel size of the imaging sensor and the focal length of the lens; z is the distance between the

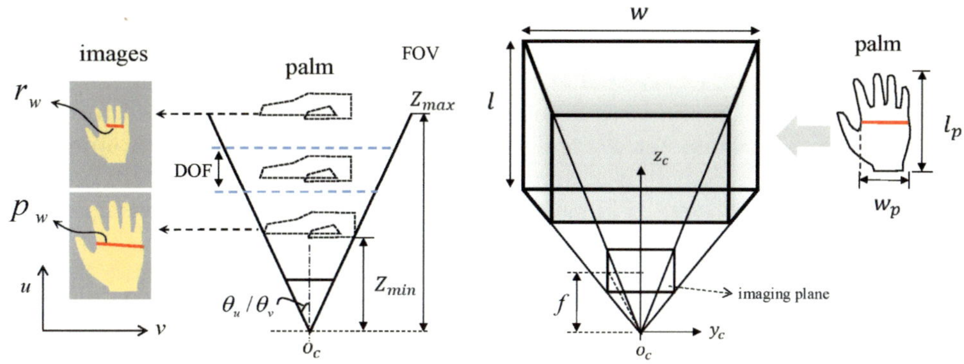

Figure 3.
Imaging model and related notations.

palm and the camera's optical center. So p_w changes according to different palm distances. Eq. (3) shows the constraints of the image palm width p_w, equivalent focal length f, palm distance z, and the palm width w_p. According to Eqs. (2) and (3), we have

$$z = f/ppi \tag{4}$$

Hence,

$$Z_{max} = f/ppi_{min} \tag{5}$$

where ppi_{min} is the minimum PPI for palmprint recognition. So, what we need to know is the relation between image PPI and system equal error rate (EER). Here, EER is an index of the system's recognition performance; lower is better. In data collection process, it is very difficult to let the users to put and hold their hands at the designed target distances, so we plan to utilize the public database to conduct simulation experiments to study the relationship between EER and PPI. In this section, database COEP [20] is selected to use, due to it is collected in a highly constrained environment. The images in it are captured by single-lens reflex camera (SLR), so they have a high signal-to-noise ratio (SNR) and very low distortions. During capturing, the user's palm is put stably on the backboard. The image resolution is sufficient to record the palmprint ridges and valleys. So we take images in COEP as the ground truth; it means they are captured with proper focus and sufficient PPI. Then the images are resized to generate palm images with different PPI. The mean PPI of a database is defined as

$$\overline{ppi} = \frac{1}{N}\sum_{i=1}^{N} ppi_i \tag{6}$$

where N is the image number of the dataset and ppi_i is the ppi value of the i-th palm image. However, in practice the captured image may contain radial and tangential distortions. The distortion parameters of the imaging model could be estimated by camera calibration [19]. Based on the imaging model, the captured image could be undistorted. Image undistortion also introduces image blur to the undistorted image. Taking this into consideration, we select four different kinds of lenses for testing, they are long-focus, standard, wide-angle, and ultrawide-angle lenses (as is shown in **Figure 4**). We use them to capture checkboard images from different views. After camera calibration, we got the corresponding intrinsic parameters. They are listed in **Table 2**. f_u and f_v are focal length along u and v directions, respectively. θ_u and θ_v are half angle of the FOV along u and v directions, respectively. k_1, k_2, and k_3 are radial distortion coefficients. p_1 and p_2 are tangential distortion factors. As is shown in **Figure 5**, the images in COEP first are distorted by the four distortion parameter sets and then undistorted by coordinates mapping and pixel interpolation based on the distortion model. The obtained images are further resized to generate different PPI palm images. According to [21], the average palm width is 84 mm for male and 74 mm for female. In [22], the average palm width is 84.18±6.81 mm for German and 82.38±11.82 mm for Chinese, and most of their subjects are male. Since palm width varies with gender, age, and race, it depends on the specific application scenarios. For simplicity, we set $\overline{w_p} \approx 80\text{mm}$ (3.15 inches) and $l_p \approx 110$ mm (4.33 inches) in our work. The original image size of COEP is 1600 × 1200. In order to delete the background area, they are cropped to size of 1280 × 960. In this experiment, we totally generate 10 datasets by image resizing; detail statistical information is listed in **Table 3**. For each palm image, using the ROI localization method proposed in [1], we can detect the tangent line of the two finger valleys, and then r_w is obtained. p_w also could be detected based on the relative coordinate system of the palm. Given a dataset, the mean p_w and mean r_w are defined as.

$$\overline{p_w} = \frac{1}{N}\sum_{i=1}^{N} p_w^i \tag{7}$$

$$\overline{r_w} = \frac{1}{N}\sum_{i=1}^{N} r_w^i \tag{8}$$

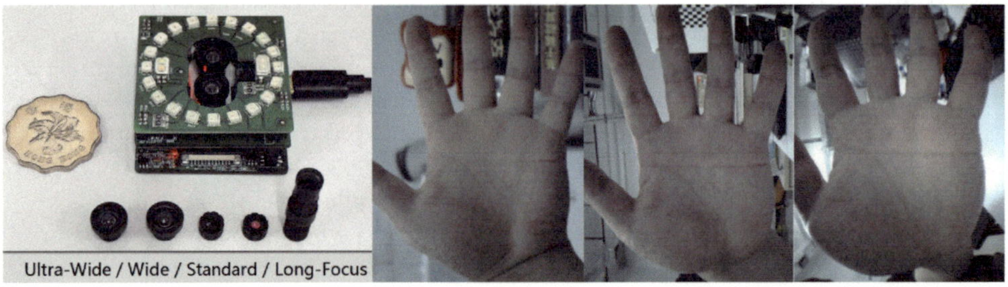

(a) (b) (c) (d)

Figure 4.
Images captured by different lenses. (a) The imaging device and different kinds of lenses. (b) An image captured by long-focus lens. (c) An image captured by standard lens. (d) An image captured by ultrawide-angle lens.

Lens	f_u	f_v	θ_u	θ_v	k_1	k_2	k_3	p_1	p_2
Long-focus	3507.05	3497.24	10.4°	7.9°	−0.37	−1.36	—	−0.0018	−0.0000
Standard	706.96	707.29	48.7°	37.5°	0.13	−0.51	—	0.0055	−0.0001
Wide-angle	435.57	436.10	72.6°	57.7°	−0.41	0.14	—	0.0014	0.0006
Ultrawide	217.19	217.99	111.7°	95.5°	0.05	−0.07	0.0105	−0.0002	−0.0018

Table 2.
The calibrated parameters of different camera lenses.

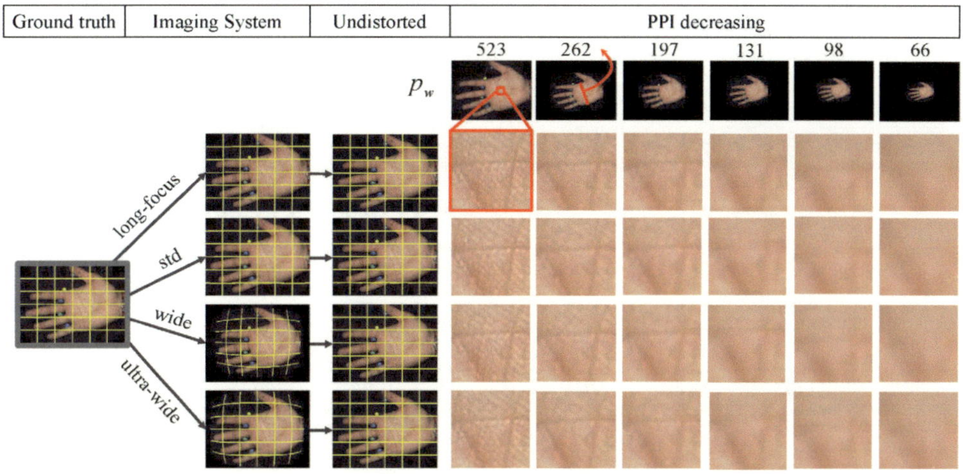

Figure 5.
Images obtained at different distances (PPI) using different distortion models.

Palm region size	1280 ×960	1120 ×840	960 ×720	800 ×600	640 ×480	480 ×360	320 ×240	240 ×180	160 ×120	80 ×60
$\overline{r_w}$	304.8	266.7	228.6	190.5	152.4	114.3	76.2	57.2	38.1	19.1
$\overline{p_w}$	524.8	459.2	393.6	328.0	262.4	196.8	131.2	98.4	65.6	32.8
$\overline{ppi}$	166.6	145.8	125.0	104.1	83.3	62.5	41.7	31.2	20.8	10.4

Table 3.
Palm region size, palm width, and corresponding $\overline{ppi}$.

where N is the image number of the dataset and p_w^i and r_w^i are the p_w and r_w values of the i-th palm image. Here, $\overline{p_w}$ is selected as the index to measure the resolution of the palm image. The sample images and corresponding enlarged local patches of the generated datasets are shown in **Figure 5**. **Table 4** describes the EERs and thresholds obtained by CompCode on different datasets. Here, $\overline{eav}$ is an index for sharpness assessment [23]. It should be noted that the sharpness level ($\overline{eav}$) obtained here has not taken the defocus-blur into consideration. It will be further studied in the next subsection. The distribution curves of $\overline{p_w}$ and corresponding EER and $\overline{eav}$ are shown in **Figure 6**. From it, we can see that the affection on image sharpness caused by undistortion is not quite obvious. Among the four cameras (as is shown in **Figure 4**), the long-focus lens obtains the highest sharpness, and wide-angle lens reaches the lowest sharpness. As to the ultrawide-angle lens, many newly designed lenses have improved their optical models to generate big distortions just in the boundary regions and small distortions in the center region. In this experiment, the wide-angle lens gains more distortions than the ultrawide-angle lens; it depends on the specific optical model the manufacturer used. Generally, the palm is put at the center of the image, so the differences between the four lenses are not large. Although the long-focus lens can provide high sharpness palm images, in real-world scenarios, the wide-angle lens is more recommended because its wide FOV provides better user experience for image capturing. As is shown in **Figure 6**, the EERs increase drastically when $\overline{p_w}$ is less than 130 pixels. So when we were selecting the imaging sensor and determining the working distance, at least we should guarantee, in the final palm image, the palm width should be large than 130 pixels; 300 pixels is recommended according to **Figure 6**.

$\overline{p_w}$	Long-focus		Standard		Wide-angle		Ultrawide	
	EER (%)	$\overline{eav}$	EER (%)	$\overline{eav}$	EER (%)	$\overline{eav}$	EER (%)	$\overline{eav}$
524.8	1.445	29.0	1.539	28.6	1.508	28.1	1.634	28.4
459.2	1.477	26.5	1.634	26.3	1.571	25.9	1.602	26.1
393.6	1.445	26.1	1.619	25.9	1.553	25.5	1.634	25.8
328.0	1.414	25.4	1.571	25.3	1.550	25.1	1.631	25.3
262.4	1.414	23.7	1.602	23.6	1.508	23.4	1.539	23.6
196.8	1.477	23.9	1.571	23.5	1.539	23.1	1.602	23.2
131.2	1.508	20.2	1.783	20.0	1.634	19.7	1.627	19.8
98.4	1.571	18.4	1.759	18.3	1.728	18.1	1.728	18.2
65.6	2.177	14.8	2.136	14.7	2.325	14.6	2.262	14.7
32.8	6.346	9.9	6.313	9.8	6.274	9.8	6.535	9.8

Table 4.
The EERs obtained from different palm width using different lens models.

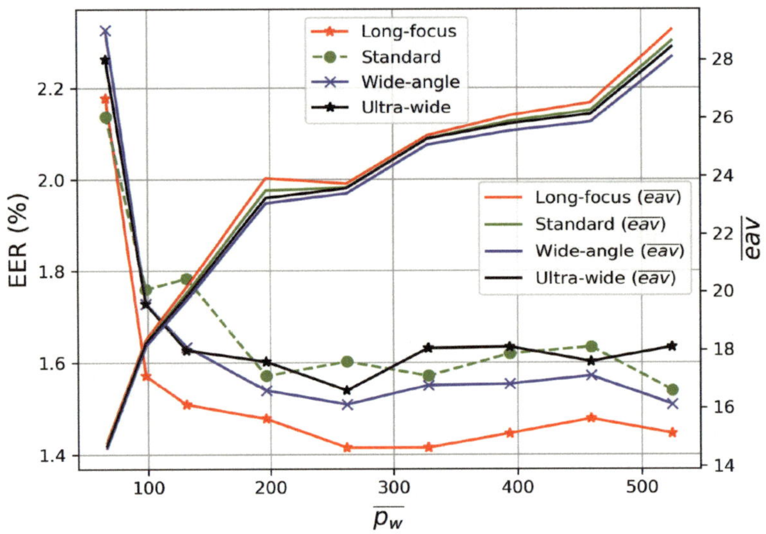

Figure 6.
The relationship between recognition performance, image sharpness, and palm width (in units of pixel).

3.2 Image sharpness range and recognition performance

In the above subsection, based on the imaging model and the capture device, we studied the relationship between palm distance, PPI, and EER. However, the hardware and the parameters of the imaging model are not always available in practice. Besides FOV, depth of focus (DOF) should be considered, since defocus-blur also will affect the final accuracy. DOF is highly related to specific applications. Our previous work [23] shows that the accuracy of palmprint recognition has a relationship with the image sharpness. Here, what we want to know is in which sharpness range the palmprint recognition accuracy is acceptable.

In this section, we try to analyze the palmprint image sharpness based on the Gaussian scale space [24]. The transform function is defined as

$$L(x,y) = I(x,y) * G(\sigma) \tag{9}$$

where x, y is the specific coordinates of the pixel and σ is the scale-coordinate. $G(\sigma)$ is the Gaussian smooth filter used for smooth the input image, and σ is its standard deviation. I is the initial image, and L is the smoothed image. So images in the scale space have different sharpness levels. As is shown in **Figure 7**, scale space function tries to generate all the potential palmprint images that may be captured in practice. In order to achieve the scale-invariant capacity, SIFT [24] tries to utilize all the information of the scale space. The method proposed in [25] is utilized here to conduct SIFT-based palmprint verifications, in which each palmprint ROI image will match against all the other images in the database. After SIFT feature extraction and matching, the random sample consensus (RANSAC) algorithm will be used to further delete the outliers. The matching between two images captured from the same palm is genuine matching, and the matching between two images captured from different palms is impostor matching. The matching number is selected as the matching score. A Gaussian image pyramid is a sampling subset of the Gaussian scale space. We wonder whether all the image layers in the Gaussian image pyramid has the same contribution to the final matches. In this experiment, once two key points from the two intra-class images are matched, the points' scales are recorded. At last, the statistical information of σ is shown in **Figure 8**. From it, we can see that the contributions of different scales are not the same; most of the distinctive local patterns only exist in some specific scales. The other layers are not discriminative. So the captured palm ROI image should not fall into those useless scale ranges. In fact, the palmprint shows different patterns at different scales. When the image is captured clearly, the palmprint consists of principal lines, wrinkles, ridges, valleys, and some minutiae points. When σ is increasing, the palmprint ROI image tends to show the spot patterns; the fine-grained ridges and valleys are smoothed and reduced to be large-scale textures. It could be seen in **Figure 1**. Different patterns have different discriminative capacities; as a result, the recognition performance changes with the image sharpness. In practice, the scale index σ corresponds to palm distance. Once the palm is moved away from the DOF of the system, the generated image suffers from defocus-blur, and the recognition performance changes.

In order to analyze the recognition performance variations, we utilize the Gaussian image pyramid to generate palmprint images at different scales. For a

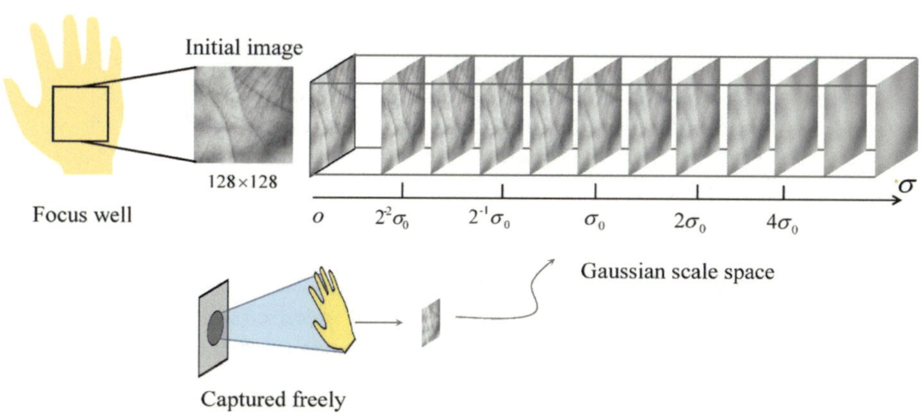

Figure 7.
The palmprint Gaussian scale space.

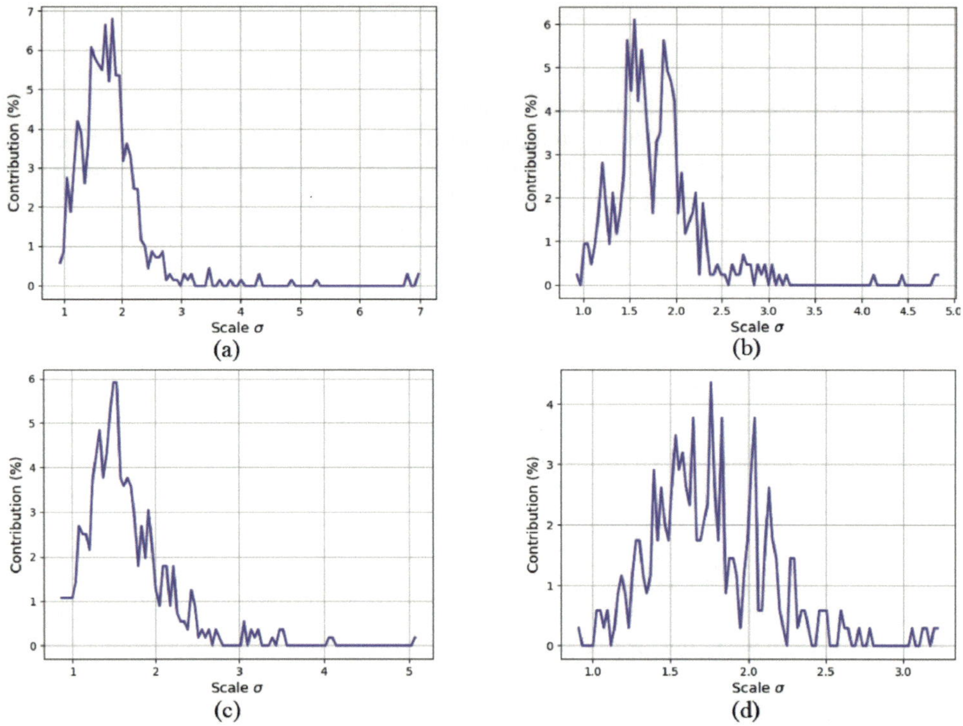

Figure 8.
Scale contributions for key point matching: (a) obtained from COEP, (b) obtained from IITD, (c) obtained from KTU, (d) obtained from GPDS.

given dataset, all the ROI images in it are filtered with Gaussian filter banks, and then 20 scaled datasets are generated. The σ used in this experiment is defined as

$$\sigma = \sigma_0 \cdot 2^{o+s/S} \tag{10}$$

$$k = 2^{1/S} \tag{11}$$

$$id = (o - o_{min}) \cdot S + s \tag{12}$$

where σ_0 is the base standard deviation; k is the step factor for increasing and decreasing σ; S is the number of intervals in each octave; o and s are octave and interval induces, respectively; and id is the image layer ID in the Gaussian scale space. o_{min} is the minimum octave index. If $o_{min} < 0$, it can generate a σ smaller than σ_0. Here, $\sigma_0 = 1.6 * k$ which is the default setting in VLfeat [26]. In this experiment, $o_{min} = -2$, $s_{min} = 0$, and $S = 4$, so the range of σ is from 0.476 to 5.709, which covers the range used in [27]. So, given one dataset, we can generate 20 datasets according to different scales. The mean EAV ($\overline{eav}$) is utilized to quantify the sharpness level of each generated dataset. **Figure 9** shows the distributions of $\overline{eav}$ and scale index σ on different publicly available palmprint databases. It shows that the sharpness level decreases almost linearly with id in the Gaussian scale space when id is smaller than 10 ($\sigma = 2.3$). Of course, the specific parameters of the curves are not the same on different databases; they are related to the database's initial sharpness level $\overline{eav}$.

The work reported in [27] shows that there exist a relationship between the recognition performance and the image sharpness. In their work, a sharpness adjustment technique is developed to improve the system EER. Different sharpness induces are tested, and EAV performs better. But only one touch-based palmprint

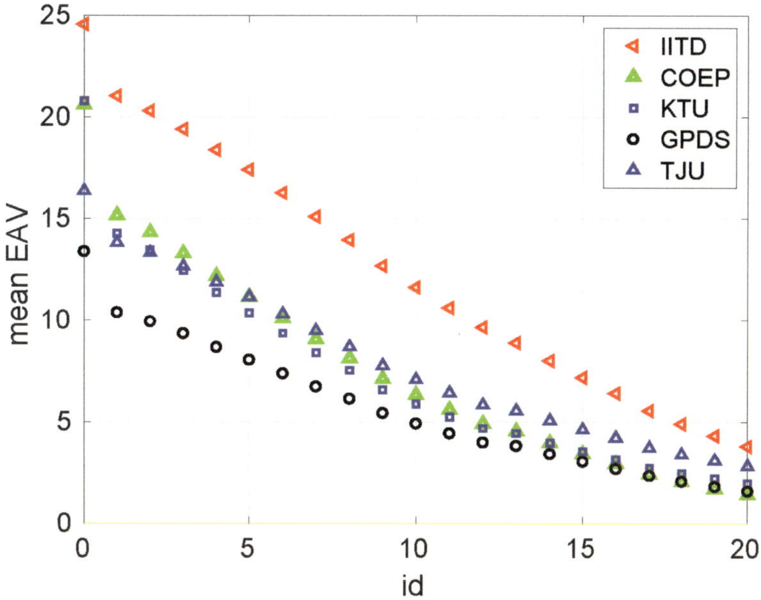

Figure 9.
The curves of $\overline{eav}$ *and corresponding scale induces on different databases.*

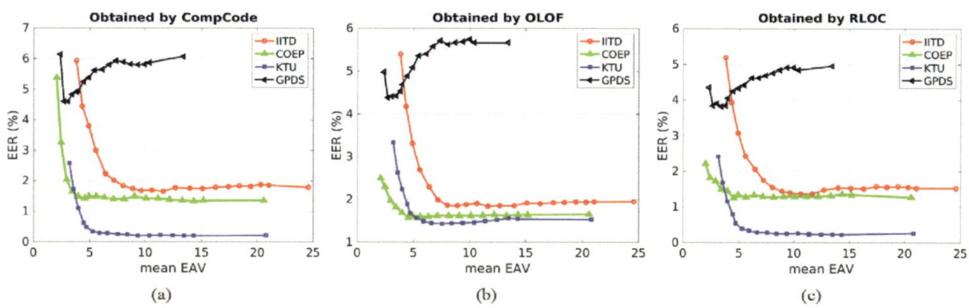

Figure 10.
The curves of EER and $\overline{eav}$ *on different databases obtained by different recognition algorithms. (a) The EER is obtained by Competitive Code. (b) The EER is obtained by OLOF. (c) The EER is obtained by RLOC.*

database is tested in their study. In order to ensure the idea is applicable on different databases, devices, and algorithms, we utilize CompCode [28], OLOF [29], and RLOC [30] to further test the recognition accuracy variations on those generated datasets. In this experiment, different databases are used including GPDS [31], IITD [32], KTU [33], and TJU [34]. **Figure 10** shows the curves of EER and corresponding $\overline{eav}$. From it we can see that the trend of GPDS is not the same with the other databases. It is because GPDS is a difficult database, which contains big illumination variations and localization errors. Hence, the recognition accuracy of this database is affected more by other factors. According to **Figure 10**, in order to guarantee the system's discriminative capacity, $\overline{eav}$ should be large than 10.

4. Conclusions

When designing a touchless palmprint recognition system, FOV and DOF are two key problems of palmprint imaging. FOV is related to image PPI, and DOF is

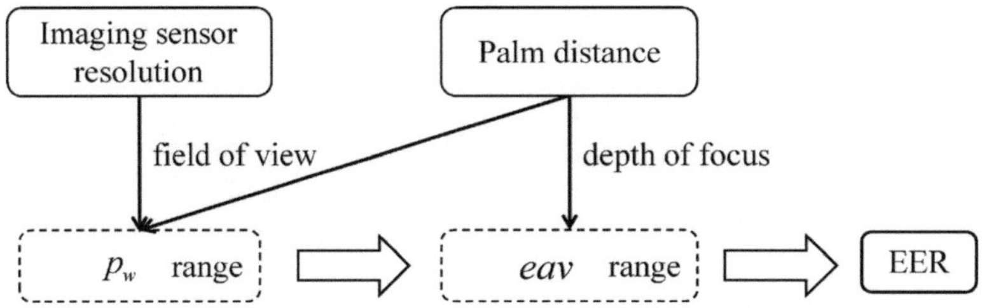

Figure 11.
The framework of this chapter.

related to image blur. **Figure 11** shows the main idea and framework of our system. In this chapter, we first studied the required image PPI for palmprint identification. Based on it, the minimum and maximum palm distances are determined in the FOV. It also provides a reference for image sensor resolution selection. Then, image blur is taken into consideration; different datasets are generated by Gaussian scale space function. The EER variation curves are obtained by different features on different databases. During the image collection process, when the palm moves out of the DOF, the sharpness of the captured image changes, so *eav* can be an index to show whether the palm is put correctly in the DOF.

Based on the findings of this research, when designing new systems, the palm width in the captured image should be larger than 300 pixels; it at least should not smaller than 130 pixels. After the system is deployed, when the user is putting his/her hand, the *eav* of the ROI image should be larger than 10. A more precise *eav* threshold should be obtained from the training dataset of the real system, because some other factors may affect the final EER distributions, such as the auto-exposure-control and auto-white-balance-control functions of the imaging sensor. But the major trends are similar. The main contribution of this work is providing some key references for system design based on image sharpness.

Acknowledgements

This work is supported in part by the NSFC under grant 61332011, in part by the Shenzhen Fundamental Research under grants JCYJ20180306172023949 and JCYJ20170412170438636, in part by the Shenzhen Institute of Artificial Intelligence and Robotics for Society.

Author details

Xu Liang[1,3]†, Zhaoqun Li[2,3]†, Jinyang Yang[1] and David Zhang[1,3,4]*

1 Harbin Institute of Technology, Shenzhen, China

2 The Chinese University of Hong Kong, Shenzhen, China

3 Shenzhen Institute of Artificial Intelligence and Robotics for Society, Shenzhen, China

4 School of Science and Engineering, The Chinese University of Hong Kong, Shenzhen, China

*Address all correspondence to: davidzhang@cuhk.edu.cn; csdzhang@comp.polyu.edu.hk

† These authors are contributed equally.

References

[1] Zhang D, Kong WK, You J, Wong M. Online palmprint identification. IEEE Transactions on Pattern Analysis and Machine Intelligence. 2003;**25**(9): 1041-1050

[2] Han Y, Sun Z, Wang F, Tan T. Palmprint recognition under unconstrained scenes. In: Proceedings of the 8th Asian Conference on Computer Vision (ACCV'07); 18-22 November 2007; Tokyo, Japan. Vol. 4844. Switzerland: Springer; LNCS(PART2). 2007. pp. 1-11

[3] Michael GKO, Connie T, Jin ATB. Touch-less palm print biometrics: Novel design and implementation. Image and Vision Computing. 2008;**26**(12): 1551-1560

[4] Zhang D, Lu G, Li W, Zhang L, Luo N. Palmprint recognition using 3-D information. IEEE Transactions on Systems, Man, and Cybernetics, Part C (Applications and Reviews). 2009; **39**(5):505-519

[5] Zhang D, Guo Z, Lu G, Zhang L, Zuo W. An online system of multispectral palmprint verification. IEEE Transactions on Instrumentation and Measurement. 2010;**59**(2):480-490

[6] Michael GKO, Connie T, Jin ATB. Design and implementation of a contactless palm print and palm vein sensor. In: Proceedings of the 11th International Conference on Control, Automation, Robotics and Vision, (ICARCV'10). 7–10 December 2010. Singapore; NewYork: IEEE; 2010. pp. 1268-1273

[7] Bu W, Zhao Q, Wu X, Tang Y, Wang K. A novel contactless multimodal biometric system based on multiple hand features. In: Proceedings of International Conference on Hand-Based Biometrics, (ICHB'11); Hong Kong, China. New York: IEEE; 2011. pp. 289-294

[8] Jia W, Hu RX, Gui J, Zhao Y, Ren XM. Palmprint recognition across different devices. Sensors. 2012;**12**(6): 7938-7964

[9] Zhao Q, Bu W, Wu X, Zhang D. Design and implementation of a contactless multiple hand feature acquisition system. In: Sensing Technologies for Global Health, Military Medicine, Disaster Response, and Environmental Monitoring II; and Biometric Technology for Human Identification IX. Vol. 8371. 2012. pp. 83711Q. Available from: https://www.spiedigitallibrary.org/conference-proceedings-of-spie/8371/1/Design-and-implementation-of-a-contactless-multiple-hand-feature-acquisition/10.1117/12.919100.full [Accessed: 03 May 2020]

[10] Nikisins O, Eglitis T, Pudzs M, Greitans M. Algorithms for a novel touchless bimodal palm biometric system. In: Proceedings of 2015 International Conference on Biometrics, (ICB'15); 19-22 May 2015; Phuket, Thailand. New York: IEEE; pp. 436-443

[11] Qu X, Zhang D, Lu G. A novel line-scan palmprint acquisition system. IEEE Transactions on Systems, Man, and Cybernetics: Systems. 2016;**46**(11): 1481-1491

[12] Qu X, Zhang D, Lu G, Guo Z. Door knob hand recognition system. IEEE Transactions on Systems, Man, and Cybernetics: Systems. 2017;**47**(11): 2870-2881

[13] Zhang L, Cheng Z, Shen Y, Wang D. Palmprint and palmvein recognition based on DCNN and a new large-scale contactless palmvein dataset. Symmetry. 2018;**10**(4):1-15

[14] Li W, Zhang D, Lu G, Luo N. A novel 3-D palmprint acquisition system. IEEE Transactions on Systems, Man, and Cybernetics - Part A. 2012;**42**(2):443-452

[15] Guo Z, Zhang D, Zhang L, Zuo W, Lu G. Empirical study of light source selection for palmprint recognition. Pattern Recognition Letters. 2011;**32**(2): 120-126

[16] Guo Z, Zhang D, Zhang L. Is white light the best illumination for palmprint recognition? In: Proceedings of International the 13th Conference on Computer Analysis of Images and Patterns (CARP'09); 2–4 September 2009; Münster, Germany. Switzerland: Springer; 2009. pp. 50-57

[17] Liang X, Zhang D, Lu G, Guo Z, Luo N. A novel multicamera system for high-speed touchless palm recognition. In: IEEE Transactions on Systems, Man, and Cybernetics: Systems. 2019. DOI: 10.1109/TSMC.2019.2898684. Available from: https://ieeexplore.ieee.org/abstract/document/8666082

[18] Tian C, Xu Y, Zuo W. Image denoising using deep CNN with batch renormalization. Neural Networks. 2020;**121**:461-473

[19] Zhang Z. A flexible new technique for camera calibration. IEEE Transactions on Pattern Analysis and Machine Intelligence. 2000;**22**(11): 1330-1334

[20] COEP database. Available from: https://www.coep.org.in/resources/coeppalmprintdatabase [Accessed: 03 May 2020]

[21] Average Hand Size For Men, Women, And Children. [Internet] Available from: https://www.theaveragebody.com/average-hand-size/ [Accessed: 09 May 2020]

[22] Rau PP, Zhang Y, Biaggi L, Engels R, Qian L, Ribjerg H. How large is your phone? A cross-cultural study of smartphone comfort perception and preference between Germans and Chinese. Procedia Manufacturing. 2015; **3**:2149-2154

[23] Zhang K, Huang D, Zhang B, Zhang D. Improving texture analysis performance in biometrics by adjusting image sharpness. Pattern Recognition. 2017;**66**:16-25

[24] Lowe D. Distinctive image features from scale-invariant keypoints. International Journal of Computer Vision. 2004;**60**(2):91-110

[25] Wu X, Zhao Q, Bu W. A SIFT-based contactless palmprint verification approach using iterative RANSAC and local palmprint descriptors. Pattern Recognition. 2014;**47**(10):3314-3326

[26] Vedaldi A, Fulkerson B. VLFeat: An Open and Portable Library of Computer Vision Algorithms. Available from: http://www.vlfeat.org [Accessed: May 03, 2020]

[27] Zhang K, Huang D, Zhang D. An optimized palmprint recognition approach based on image sharpness. Pattern Recognition Letters. 2017;**85**:65-71

[28] Kong A, Zhang D. Competitive coding scheme for palmprint verification. In: Proceedings of the 17th International Conference on Pattern Recognition (ICPR'04); 23–26 August 2004. Cambridge, UK. New York: IEEE; 2004. pp. 520-523

[29] Sun Z, Tan T, Wang Y, Li SZ. Ordinal palmprint representation for personal identification [represention read representation]. In: Proceedings of the IEEE Computer Society Conference on Computer Vision and Pattern Recognition (CVPR'05); 20–26 June 2005. San Diego, CA. New York: IEEE; 2005. pp. 279-284

[30] Jia W, Huang D, Zhang D. Palmprint verification based on robust line orientation code. Pattern Recognition. 2008;**41**(5):1504-1513

[31] GPDS database. Available from: www.gpds.ulpgc.es/downloadnew/download

[32] IITD database. Available from: https://www4.comp.polyu.edu.hk/~csajaykr/IITD/Database_Palm.htm [Accessed: 03 May 2020]

[33] KTU database. Available from: https://ceng2.ktu.edu.tr/~cvpr/contactlessPalmDB.htm [Accessed: 03 May 2020]

[34] TJU database. Available from: https://sse.tongji.edu.cn/linzhang/contactlesspalm/index.htm [Accessed: 03 May 2020]

Chapter 6

Transfer Learning of Pre-Trained CNN Models for Fingerprint Liveness Detection

Hussein Samma and Shahrel Azmin Suandi

Abstract

Machine learning experts expected that transfer learning will be the next research frontier. Indeed, in the era of deep learning and big data, there are many powerful pre-trained CNN models that have been deployed. Therefore, using the concept of transfer learning, these pre-trained CNN models could be re-trained to tackle a new pattern recognition problem. As such, this work is aiming to investigate the application of transferred VGG19-based CNN model to solve the problem of fingerprint liveness recognition. In particular, the transferred VGG19-based CNN model will be modified, re-trained, and finely tuned to recognize real and fake fingerprint images. Moreover, different architecture of the transferred VGG19-based CNN model has examined including shallow model, medium model, and deep model. To assess the performances of each architecture, LivDet2009 database was employed. Reported results indicated that the best recognition rate was achieved from shallow VGG19-based CNN model with 92% accuracy.

Keywords: transfer learning, pre-trained CNN model, VGG model, deep learning, fingerprint, liveness detection

1. Introduction

Recently, deep CNN models have been successfully applied for many pattern recognition problems such as human facial expression recognition [1], vehicle detection [2], and lung diseases diagnosis [3]. The application of CNN models for fake fingerprint recognition was investigated by Nogueira et al. [4]. Particularly, they have studied the effectiveness of different schemes including Local Binary Patterns (LBP), SVM, VGG, and Alexnet model. These discussed models were evaluated using the dataset of liveness detection competition for the years of 2009, 2011, and 2013. The outcomes of Average Classification Error (ACE) measure showed that the best accuracy of was reported by VGG-based deep model was 3.4. Further Anti-spoofing approach for fingerprint recognition was conducted by Uliyan [5]. They have presented deep Restricted Boltzmann Machines (RBM) to encode and represent the features. Then, KNN classifier was used to classify the input pattern as real or fake case. To assess the performances of RBM-KNN model in [5], LivDet dataset was used. Reported results showed that 3.6 ACE value was achieved on LivDet 2013 benchmark images.

An incremental learning approach was given by Kho et al. [6]. The key idea is that an ensemble of SVM classifiers was constructed using boosting technique. Specifically, each base classifier in the ensemble model was trained with different subsets of the given training set. For feature extraction, three different types of hand-crafted features were utilized namely LPQ, LBP, and BSIF. Experimental results indicated that the presented ensemble model outperforms single SVM classifier. In addition, they have investigated the performances of CNN as a feature extractor with ensemble model as a classifier. The outcomes show the superiority of deep CNN features against the classical hand-crafted features, that is, LPQ, LBP, and BSIF. A recent deep CNN-based approach was discussed by Fei et al. [7]. In their work VGG19, Alexnet and Mobilenet CNN models were employed. Their models were retrained with LiveDet2013 and LiveDet2015 images. The outcome indicated that the best accuracy performance was achieved from VGG19 among other CNN-based models.

Nowadays, transfer learning becomes a promising technique that could be applied to utilize and reuse a powerful pre-trained CNN models to handle different pattern problems. For example, a transferred CNN models was applied for the recognition of brain tumors [8], wildfire detection [9], pneumonia diagnosis [10], seizure classification [11], remote sensing image retrieval [12], and bearing fault detection [13]. Nevertheless, the idea of transfer learning of a pre-trained CNN network is considered as a new and has not been widely studied for liveness detection. As such, this work is aiming to investigate transferring of various architectures of VGG19 CNN model to handle the problem of liveness detection. The remaining part of this chapter is organized as follows. The proposed transferred model is explained in Section 2. A series of experiments has been conducted to evaluate the effectiveness of the proposed approach is given in Section 3. A summary of the research findings and conclusions of this study is presented in Section 4.

2. Architecture of pre-trained VGG19 CNN model

The basic architecture of VGG19 CNN model is given in **Figure 1**. As can be seen that VGG network contains four different types of layers namely convolution layer, max-pool layer, fully connected layer (FC), and soft-max classification layer.

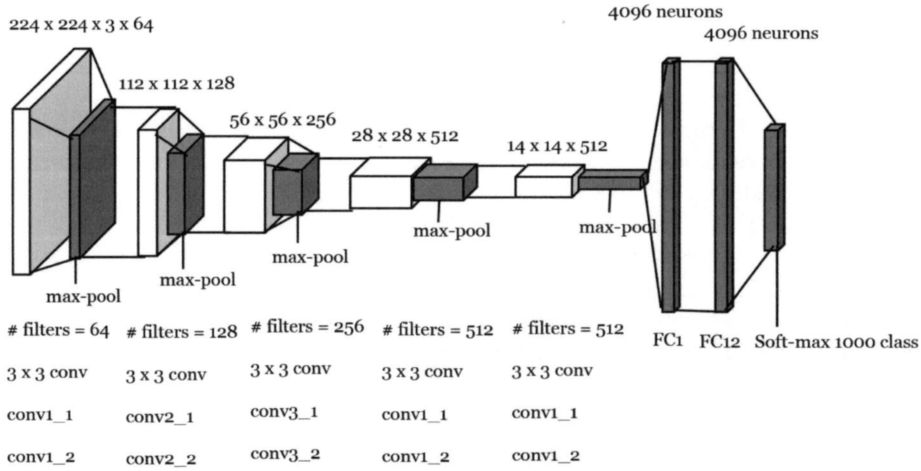

Figure 1.
VGG19 architecture.

The main aim of convolution layer is to perform convolution operation of a pre-trained filter with the input image. As indicated in **Figure 1**, the input image size is 224 × 224 × 3 and the first layer consists of 64 filter of size 3 × 3. Going deeper into VGG, the number of convolution filters has been increased from 64 to 512 as shown in **Figure 1**.

Max-pooling layer in VGG19 is used to reduce the dimensionality of input data. In particular, a sliding window of size 2 × 2 has been used for computing the max value in the sliding box which represents the reduced data. As such, after applying max-pooling operation, the image of size 224 × 224 will be reduced to half of its size and becomes 112 × 112. So, these CNN operations, that is, convolution and max-pooling are repeated until the final image size becomes 14 × 14 as shown in **Figure 1**. After that, a flattening operation is applied to reshape the data from 14 × 14 × 15 to be as 1-D vector of size 4096.

Fully connected layers in VGG19 will take an input 1-D vector of size 4096 and feed it to a fully connected neurons of size 4096. It should be noted that VGG19 contains two consecutive FC layers with the same size as shown in **Figure 1**. Finally, soft-max classifier is used to perform the task of classification. Therefore, the input image will be classified as one of the 1000 different classes which are car, dog, etc.

3. Transfer learning of pre-trained VGG-19 CNN model

The basic idea of transfer learning is to employ a pre-trained network such as VGG19, then, to perform replacement for the last layer, that is, soft-max classifier.

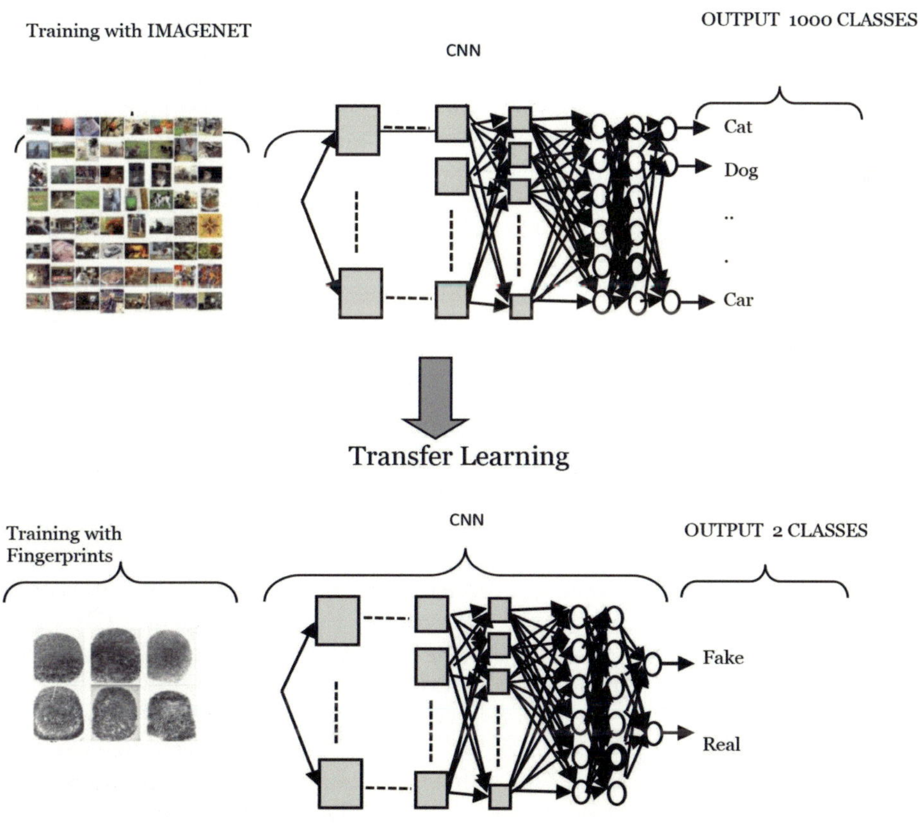

Figure 2.
Transfer learning of VGG19 for fingerprint liveness detection.

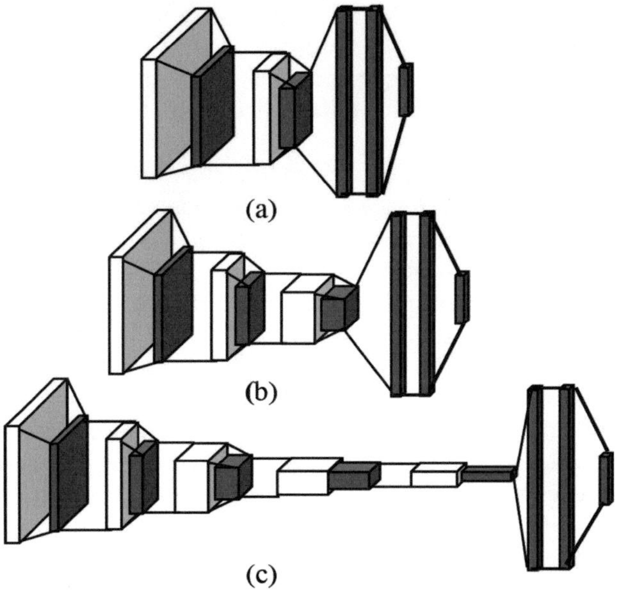

Figure 3.
Different architecture of transferred VGG19 CNN model (a) shallow (b) medium, and (c) deep.

The new classification layer will be set according to the number of classes in the problem that need to be tackled. Finally, the model will be re-trained with a new training set. This idea is described in **Figure 2**.

In this study, the performance of three different architectures of VGG19 will be investigated. The transferred models include shallow, medium, and deep model as shown in **Figure 3**. For example, shallow VGG19-based CNN model contains the first and second block of VGG19. In addition, soft-max classifier has been replaced with a new classifier with two classes, that is, neurons. One neuron of soft-max is used to recognize and give probability of fake fingerprints meanwhile the second neuron is used for recognizing real fingerprints. It should be noted that the architecture of deep VGG19 CNN model contains the whole layers except the classification layer which replaced with two neurons as explained previously.

4. Experimental analysis

This study uses the database of LivDet2009 Database [14]. A few samples for real and fake images are shown in **Figure 4**. As described in [14], fake images were collected from a cloned fingerprint using silicon material. The total number of images used in this analysis was 1040 images for training and 2953 images for testing purposes.

The conducted analysis examined three different types of VGG architecture which are shallow, medium, and deep CNN model. Besides that, a new CNN model has been crated from scratch with the same architecture of shallow model. Each CNN model in this experiment was trained using the same training set. **Table 1** shows the outcomes for each model. As can be seen from the reported results that created CNN from scratch produced the worst performances in terms of all examined measures. This is due to lack of number of training images which usually required for building deep CNN models. On the other hand, the transferred shallow VGG19-based CNN model was able to achieve the best performances in terms of accuracy, precision, recall, and F1 score. Deep CNN model achieved the lowest performances among the transferred models because it lacks for generalization as compared with shallow model.

Additional analysis was conducted by computing the confusion matrix for each model as reported in **Tables 2–5**. As shown from the results, the best true positive rate (TP) was achieved from shallow model. Specifically, the shallow model was able to correctly classify a total of 1308 cases with only 165 missing cases. In addition, shallow model reported the minimum false alarms with only 70 cases as given in **Table 3**. This is due to the benefit of transfer learning and generalization ability as compared with deep CNN models.

Further analysis was conducted by computing the receiver operating characteristic curve (ROC) for each studied model. ROC is shown in **Figure 5**, and the plotted curves show a very close results achieved from shallow and medium model. The worst performance was produced by a CNN model created from scratch as given in **Figure 5**.

Finally, the area under the curve (AUC) measure was computed for each model as given in **Table 6**. As can be shown that AUC value resulted from the transferred models outperform the outcomes of CNN model created from a scratch. This implies that transfer learning of a pre-trained models represent a good alterative to be used instead of building a new CNN model from a scratch which required a huge training data.

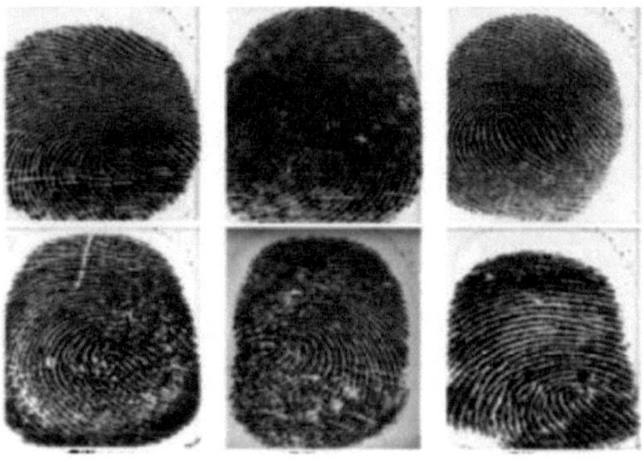

Figure 4.
Fingerprint examples from LivDet2009 database [14], real cases (top line), and fake cases (bottom line).

Approach	CNN from scratch	Transferred VGG19		
		Shallow	Medium	Deep
Accuracy (%)	69.45	92.04	86.99	81.44
Precision (%)	62.82	89.52	80.68	88.89
Recall (%)	95.67	95.27	97.36	71.96
F1 score (%)	75.84	93.27	92.31	79.54

Table 1.
Fake fingerprint recognition results.

	Predicted class	
Actual class	TP = 635	FN = 838
	FP = 64	TN =1416

Table 2.
Confusion matrix for CNN from scratch.

		Predicted class	
Actual class		TP = 1308	FN = 165
		FP = 70	TN = 1410

Table 3.
Confusion matrix for transferred VGG19 shallow.

		Predicted class	
Actual class		TP = 1128	FN = 345
		FP = 39	TN = 1441

Table 4.
Confusion matrix for transferred VGG19 medium.

		Predicted class	
Actual class		TP = 1340	FN = 133
		FP = 415	TN =1065

Table 5.
Confusion matrix for transferred VGG19 deep.

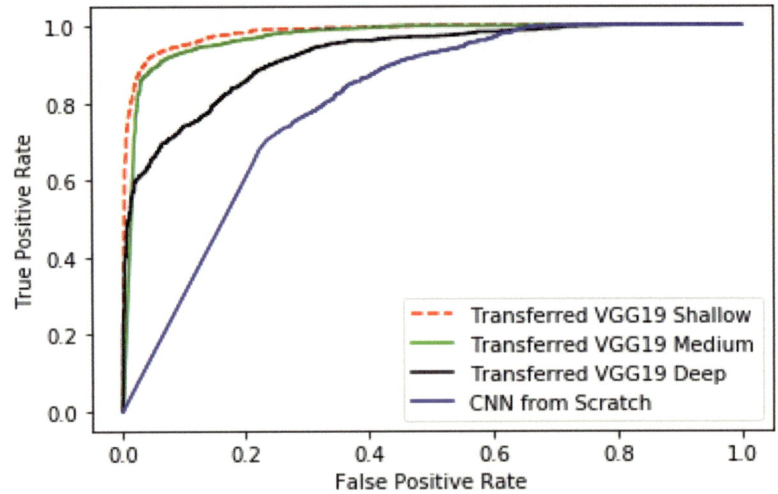

Figure 5.
Results of receiver operating characteristic (ROC).

Approach		ROC AUC
CNN from scratch		0.796899
Transferred VGG19	Shallow	0.982223
	Medium	0.969411
	Deep	0.920636

Table 6.
Fake fingerprint recognition results.

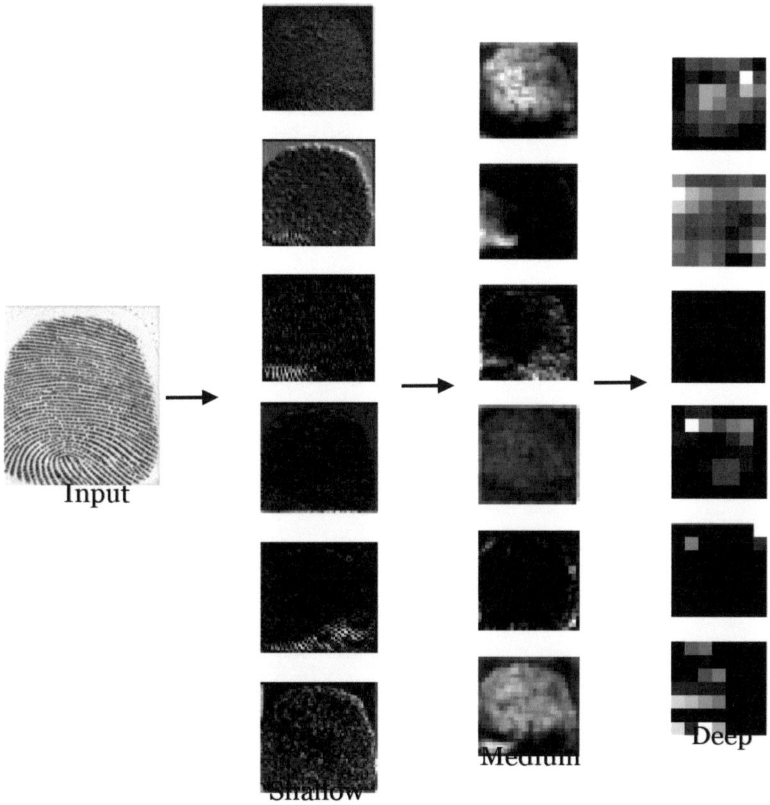

Figure 6.
Visualizing intermediate layers of the transferred VGG19-based models, that is, shallow, medium, and deep model.

Figure 6 visualizes the intermediate layers of the transferred VGG19 model for the three studied architectures, that is, shallow, medium, and deep model. As can be seen in **Figure 6** that at deep layers, the fine details of fingerprint are disappear. This is due to max-pooling operations which shrink down image size. This implies that shallow and intermediate layers produce better recognition results owing to keeping the content and details of the convolved input image as shown in **Figure 3**.

5. Conclusion

This chapter discusses the idea of transfer learning technique of a pre-trained VGG19 model to handle the problem of liveness detection of fingerprint images. A total of three different architectures of VGG19 were examined in this chapter. These architectures include shallow, medium, and deep CNN model. The reported results confirmed the performances of the transferred VGG19 models as compared with a CNN model created from scratch. Among the transferred VGG19 models, shallow model shows the best performances in terms of accuracy, precision, recall, and F1 score.

Conflict of interest

Authors declare no conflict of interest.

Author details

Hussein Samma[1*] and Shahrel Azmin Suandi[2]

1 Department of Computer Programming, Faculty of Education – Shabwa, University of Aden, Aden, Republic of Yemen

2 Intelligent Biometric Group, School of Electrical and Electronics Engineering, Universiti Sains Malaysia, Nibong Tebal, Malaysia

*Address all correspondence to: hussein.samma@yahoo.com

References

[1] Gan Y. Facial expression recognition using convolutional neural network. In: Proceedings of the 2nd International Conference on Vision, Image and Signal Processing. ACM; 2018

[2] Hu X et al. SINet: A scale-insensitive convolutional neural network for fast vehicle detection. IEEE Transactions on Intelligent Transportation Systems. 2018;**20**(3):1010-1019

[3] Anthimopoulos M et al. Lung pattern classification for interstitial lung diseases using a deep convolutional neural network. IEEE Transactions on Medical Imaging. 2016;**35**(5):1207-1216

[4] Nogueira RF, de Alencar Lotufo R, Machado RC. Fingerprint liveness detection using convolutional neural networks. IEEE Transactions on Information Forensics and Security. 2016;**11**(6):1206-1213

[5] Uliyan DM, Sadeghi S, Jalab HA. Anti-spoofing method for fingerprint recognition using patch based deep learning machine. Engineering Science and Technology, an International Journal. 2019;**23**(2):264-273

[6] Kho JB et al. An incremental learning method for spoof fingerprint detection. Expert Systems with Applications. 2019;**116**:52-64

[7] Fei J et al. Adversarial attacks on fingerprint liveness detection. EURASIP Journal on Image and Video Processing. 2020;**2020**(1):1

[8] Rehman A et al. A deep learning-based framework for automatic brain tumors classification using transfer learning. Circuits, Systems, and Signal Processing. 2020;**39**(2):757-775

[9] Sousa MJ, Moutinho A, Almeida M. Wildfire detection using transfer learning on augmented datasets. Expert Systems with Applications. 2020;**142**:112975

[10] Wu Z et al. An adaptive deep transfer learning method for bearing fault diagnosis. Measurement. 2020;**151**:107227

[11] Raghu S et al. EEG based multi-class seizure type classification using convolutional neural network and transfer learning. Neural Networks. 2020;**124**:202-212

[12] Liu Y et al. Similarity-based unsupervised deep transfer learning for remote sensing image retrieval. IEEE Transactions on Geoscience and Remote Sensing. 2020. pp: 1-18(Early Access)

[13] Mao W et al. Online detection for bearing incipient fault based on deep transfer learning. Measurement. 2020;**152**:107278

[14] Marcialis GL et al. First international fingerprint liveness detection competition—LivDet 2009. In: International Conference on Image Analysis and Processing. Springer; 2009

Chapter 7

Assessment Methods of Cognitive Ability of Human Brains for Inborn Intelligence Potential Using Pattern Recognitions

Rohit Raja, Hiral Raja, RajKumar Patra, Kamal Mehta, Akanksha Gupta and Kunta Ramya Laxmi

Abstract

This research aims to examine the scientific study related to fingerprint patterns and brains lobes. Generally, this method is used to find and develop the inborn potential and personality especially of children. Every person is having inborn potential and personality, which will help us to analyze strength and weakness. The present work is based only on the analysis and used as a reference for scientific research in the field of Galtian and statistical study conducted based on the fingerprint processing. Human brain is divided into two parts, left hemispheres and right hemispheres. Fingers of right hand represent the functions of left brain and fingers of left hand represent the functions of right brain. Human brain is divided into 10 lobes and each lobe is related with each finger. Each lobe represents different intelligences. A detailed analysis of the fingerprint would help the researchers to find the inborn talents. It will provide them the most appropriate learning habits from young age and improve learning ability effectively. The vital factor of an individual's intelligence is determined by neural network connection of brain cells. Cognitive science is the scientific study that will help you to know about yourself.

Keywords: Galtian characteristic, pattern recognition, neural network, cognitive cell, intelligence

1. Introduction

Clinical specialists, through tracking, recording, comparison, induction, and also professional trials showed that fingerprints deliver the exact evaluation of a character's innate talents. The evaluation gadget opinions the distribution of mind's understanding ability and also allotment of cerebral characteristic of a selected and additionally elements suitable statistical document of person's innate intelligence. Thus, it permits increase of the man or woman in international of leading information.

This can help the guide/mentor to recognize the inherent qualities and powerful conversation mode of the man or woman. It will provide the maximum right discovering behaviors from the young age, which would be over and above to finding out capabilities. It can likewise help the guide/mentor to recognize the development of more than one intelligence and also uncover the opportunity of the man or woman. Therefore, it will be very easy to decorate their vulnerable factor through the proposed method in a good way to accomplish ordinary development.

Last but not least we would really like to carry the message to all guides/mentors in addition to people that the aim of this test is to permit you to absolutely recognize and also appreciate precise differences of anyone and additionally provide education and mastering and/or schooling for this reason.

Our thoughts are separated into hemispheres, left and right. Each hemisphere of the mind has its very own resilience. Fingers of right hand constitute the features of left mind as well as arms of left hand represent the features of right brain. Each intelligence has its very own weight age. Overall distribution of intelligences' portion can be 100%.

In this chapter, authors applied bifurcation, termination, and neural network for feature extraction and got 90.06% accuracy for identifying an authorized person with the help of a proposed figureprint recognition method [1]. For human face recognition from side view of face, authors used Manhattan distance and support vector machine of artificial neural network, along with front view analysis and achieved up to 95.3528% accuracy in their work [2]. In the chapter, authors compared the performance of different biometric technologies like fingerprint, hand geometry, key stroke, etc. on the basis of EER, FAR, and FRP. The chapter is based on standardized fingerprint model for fingerprint matching. The author used mean images and genetic algorithm. Transformation is also used for synthesizing a fingerprint [3].

In this chapter, authors had thrown the light on the many preexisting methods and techniques of fingerprint recognition system. All four stages of fingerprint recognition system were elaborated briefly. Database related to fingerprint recognition had shown with characteristics [4]. For fingerprint recognition, the popular technique is "Euclidean distance" and "neural network classifier", whereas for preprocessing of images, "histogram equalization" and "fast Fourier transformation" are used. The result of this work was significantly better than the previous work [5].

In the chapter, authors first developed a CNN framework for more hygiene and accurate contactless fingerprint recognition and this work also helped to alleviate spoofing of fingerprint and shown much greater security than the preexisting methods [6]. In the chapter, authors used color coding scheme, Sobel and Canny method, HSV histogram, edge detector method, and Corel-1 K dataset for detection of color object [7]. In this chapter, authors gave a thorough knowledge of fingerprint recognition and also proposed a secured fingerprint recognition payment system [8].

This chapter defined many aspects, methods, and techniques like Gabor filter, FFT, minimum distance classifier, histogram equalization, fusion and context switching framework, etc. for fingerprint-based identification system [9].

In this chapter, authors applied LGXP and ANN techniques for face recognition to handle variation in human face due to change in pose, illumination condition, viewing direction, and expression of different ages [10]. In this fingerprint-based biometric review paper, authors briefly discussed different attacks, and compared different existing methods of biometric cryptosystem, cancellable biometrics, etc. for fingerprint template protection [11]. The most widely used biometrics is fingerprint technology. The fingerprint is a pattern of ridges and valleys present on the surface of a fingertip [12]. The finger ridge configurations do not change throughout the life of an individual, except in case of accidents such as burns or cuts on the fingertips. The fingerprints are so unique that two identical twins have different fingerprints [13]. Matching accuracy using fingerprint is very high as compared to other biometrical traits. Initially the fingerprint technology of biometric identification is used for forensics and criminal investigation (**Figure 1**) [14].

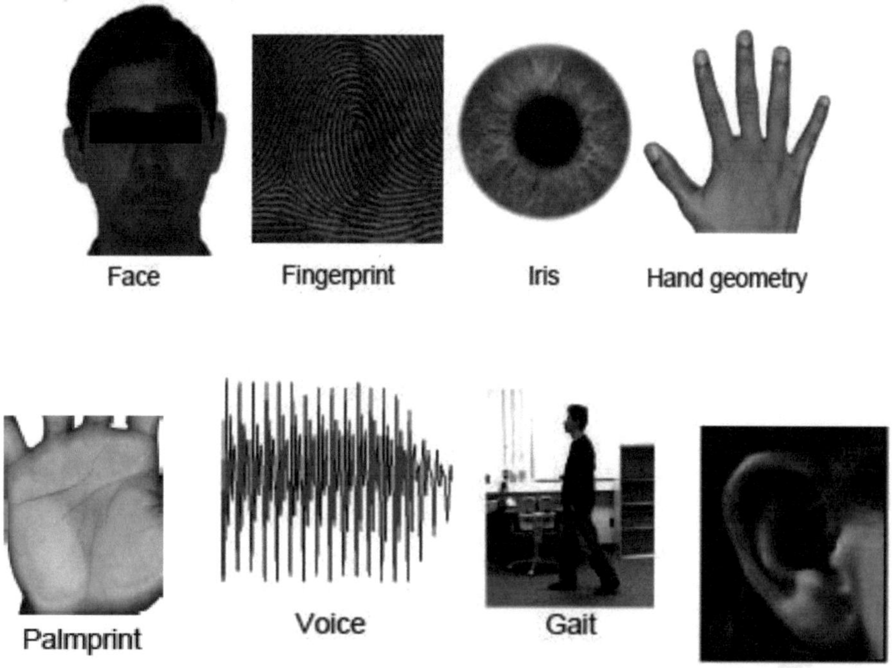

Figure 1.
Examples of body traits that can be used for biometric recognition (courtesy: http://images.google.co.in).

2. Proposed methodology

Various pattern kinds will definitely display the flow of different worth. In ordinary situations for the general public of people, the worth will truly range from 8 to 30. If the worth is high, it indicates that the function of the thoughts cortex assignment stage is high. Everybody has the capability for selected natural pinnacle traits. With boosting and locating out, it is easy to turn out to be being a better human. Support the thought that every of us can absolutely stimulate our viable and might accomplish first-rate future (**Figures 2–4**).

Inborn intelligence potential:

Low potential
Average potential
Good potential
Very good potential
Excellent potential
Hyper active

2.1 Left brain

Analytical mind is more likely to exhibit self-awareness, logical thinking, language & grammar, curiosity, and love [15]. They individuals are usually desirable in teachers. They have convergent reasoning and can deliver their power and also emphasis at one factor. They prefer to respond to Spoken guidelines. They want to fix the issues by searching at the parts of points [16]. They are in a position situate the difference transgression comparable points speedy. They are an awful lot more supposed and also based. Prefer more than one option checks. They have

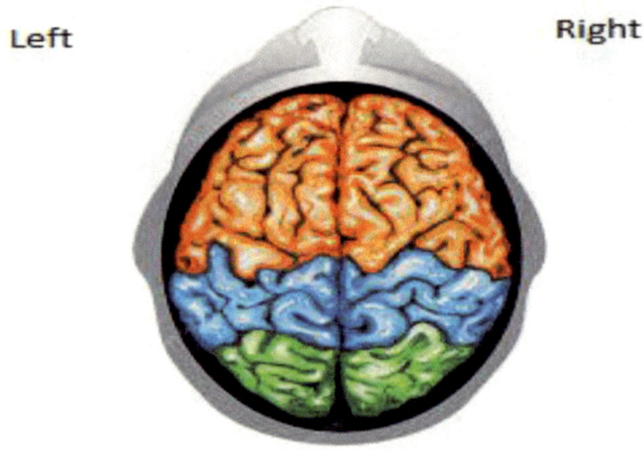

Figure 2.
Left and right part of brain (https://www.google.com/).

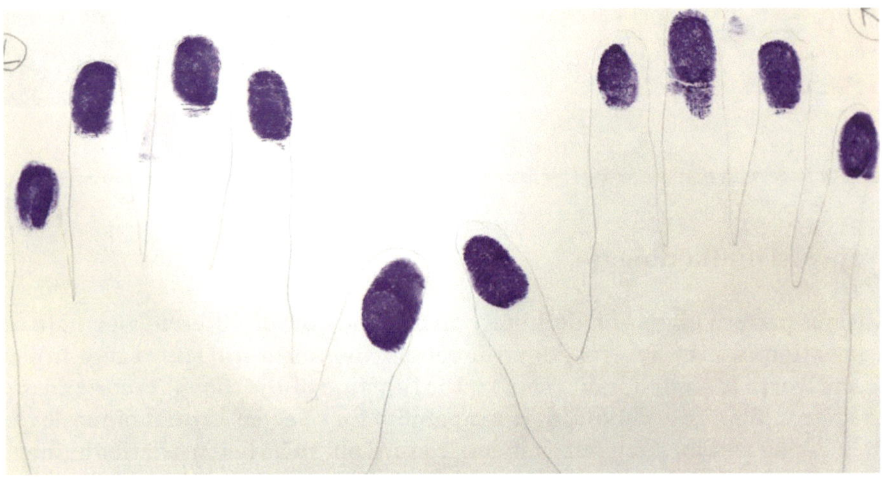

Figure 3.
Fingerprint of left and right hand of 5 years child.

the functionality to manipulate their feeling sand emotions. They like foundation, problems, word developing, problem resolving, crosswords, and so forth [17].

They include coping with the problems via checking out the problem usual. They are able draw the whole image in their thoughts quick. They are extra intuitive and paintings upon sensations. Their feelings and emotions do not have any limitations, and that they typically seem [18].

2.2 Right brain

Creative mind is tons more inclined toward social talents, creativity, gross electric powered motor competencies obligations, tune, sun shades, photos, dance, art, rhythms, appearing, paint, modeling, style, outside sporting sports, and so on. They are generally brilliant in extracurricular sports, generally creative ones. They have a tendency to throw the dismiss of window [19]. They have specific reasoning that incorporates creativeness and also thoughts. As properly as they are commonly misplaced in their personal ideas, thoughts and global. They may be actually present, psychologically lacking. Right mind individuals choose to answer to proven path (**Figure 5**) [20].

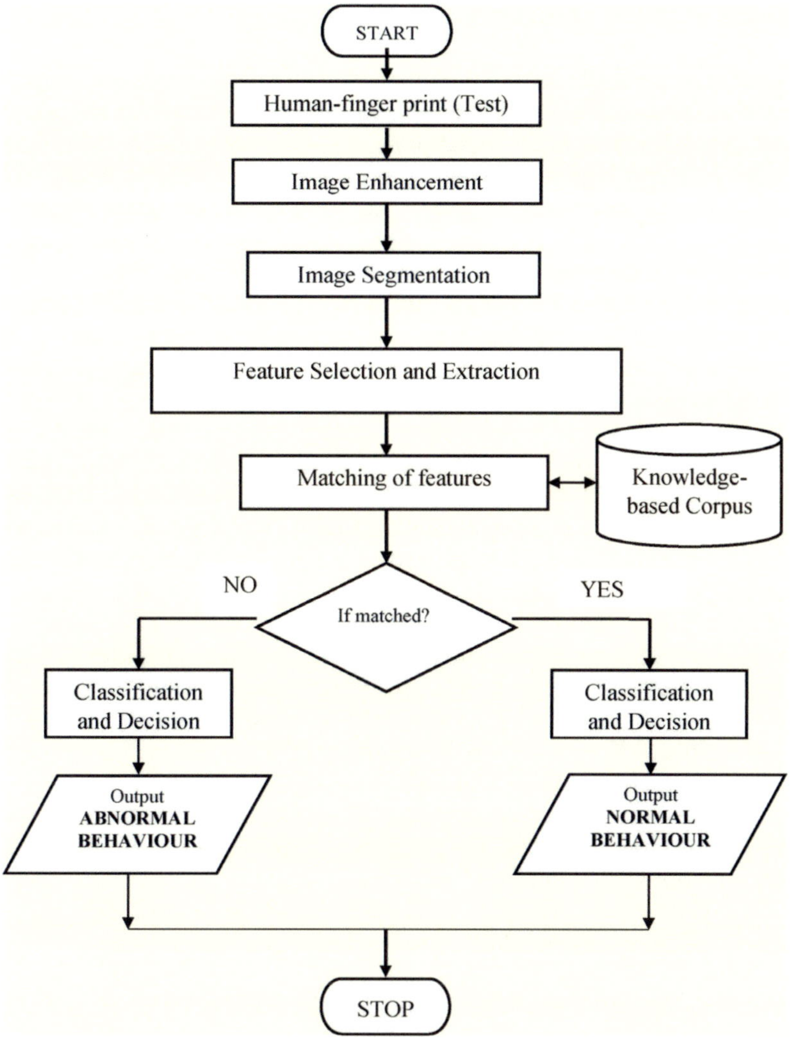

Figure 4.
Workflow diagram.

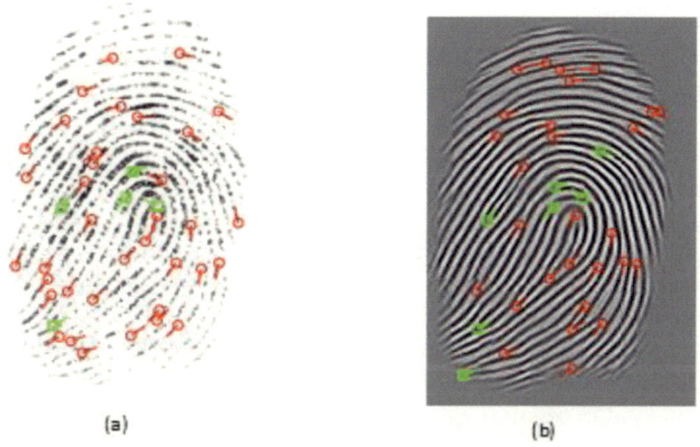

Figure 5.
Feature extraction from fingerprint.

3. Functions of brain of different sections

The brain is responsible for personality and characteristics. Impulse control, capability to evaluate social situations, socializations, spontaneity, capacity to override as well as subdue unwanted social practices, movements. Cognitive features (Exec Features) [21]: judgment, thinking, problem addressing, planning, social capabilities, control, abstract reasoning, imagination, and initiative responsibilities that need the assimilation of data in time, capability to decide similarities and differences between matters or activities, and mental features [22].

The brain is responsible for creative thinking and visualization. Abstract concept, hassle solving, summary thinking, language tasks of math, thinking, coping with phrases as well as grammar syntax, visualization, creativeness, and ideas in addition to principles formation [23].

The brain is responsible for processing auditory information. They differentiate differences in sound, pitch, and also quantity and set up their importance. The proper temporal lobe is in rate of musical appreciation, even as the left temporal lobe is liable for the expertise of speech. Left temporal lesions lead to damaged reminiscence for verbal product. Right aspect lesions lead to impaired recall of nonverbal product, consisting of track [24].

The brain is responsible for processing visual information. They process information about objects, colors, motion distance, words, signs identification of objects, and symbols [25]. Responsible for spatial awareness and for processing and analyzing sensory stimuli. They play vital roles in incorporating sensory info from several detects in addition to within the manager of things. Portions of the parietal lobes are covered with visible-spatial potential [26].

- Rational thinking, planning, coordinating, controlling, executing achievement, self-motivation, and self-awareness.
- Leadership, interpersonal skill, creativity, and goal visualization.
- Self-esteem, intuition, and the ability to understand others point of view.
- Logical reasoning, computation process, analytical skills, and conceptual understanding.
- Numeric, grammar syntax, and cause and effect relationships.
- Imagination, idea formation, visualization, 3D recognition, visual spatial ability, and hand-eye co-ordination.
- Fine motor skills, action identification and understanding, finger control, and control of body movements.
- Gross motor skills, body movement and sensory information, and eyes body co-ordination.
- Language ability, language understanding, and audio identification.
- Ability and syntax of language.
- Tone understanding, sound and voice understanding, music, emotions, and feelings.

- Visual identification, interpretation, reading, observation, image appreciation, and recognition of shapes and colors.
- Visualization, visual appreciation, art, and esthetic.
- Sense understanding of maps, visuals, graphical, and communications [27].

Further mind is cut up into two components, left brain in addition to right brain. Left brain controls a great aspect of the body and vice-versa. Science has showed that within the very identical wattle, left and right brain do different precise obligations. So, mind has 10 booths, 5 left and 5 proper; every compartment has info and pre-detailed feature [28]. Additionally, our brain has approximately one hundred billion Neuron cells, which are separated in arbitrary order into those 10 areas. It is hard that two people have very identical nerve cellular distribution [29]. One could sincerely want to do that place's paintings, in which the nerve cell count number is a lot greater. He will in reality revel in that paintings and will simply discover it clean. It will actually be longevity region. One could despise to try this compartment's work, in which nerve mobile dependency is a lot less. He will now not adore it and will without a doubt locate it difficult to do. It will in reality be a susceptible region [30].

3.1 Verbal intelligence

Preferences: write, read, tell stories, talk, memorize, work at solving puzzles, etc. Learns through: hearing and seeing words, speaking, reading, writing, discussing, debating, etc. Needs: books, tapes, papers, diaries, writing tools, dialog, discussion, debates, stories, etc. [31].

Activity involve in

- Most in all likelihood to concerts or musicals.
- Establish a collection of preferred musical recordings in addition to pay attention to them on an everyday foundation.
- Join a community choir.
- Take legit music training in a specific tool.
- Work with a song's specialist.
- Spend 1 h every week taking note of an ordinary design of songs (jazz, the United States of America; western, classical, people, international; or other categories).
- Establish an ordinary household sing-alongside time.
- Purchase an electronic keyboard and also find out honest tunes in addition to chords.
- Purchase percussion contraptions at a plaything shop and play them in rhythm to historical past track.
- Take a course in song appreciation or songs concept at a local institute.

- Read songs objection in papers and courses.
- Purchase contemporary gadgets (MIDI interface, computer machine software program) with a purpose to definitely (**Figures 6** and 7; **Table 1**) [32].

Figure 6.
Graphical representation of EQ, IQ, AQ, CQ.

	Value in Percentage
Dance	10
Instrumental Music	9.56
Horse Riding	5
Foreign Language	7.03
Painting	9.88
Singing Acting/Drama Swimming	7.34
Chess Snooker	8

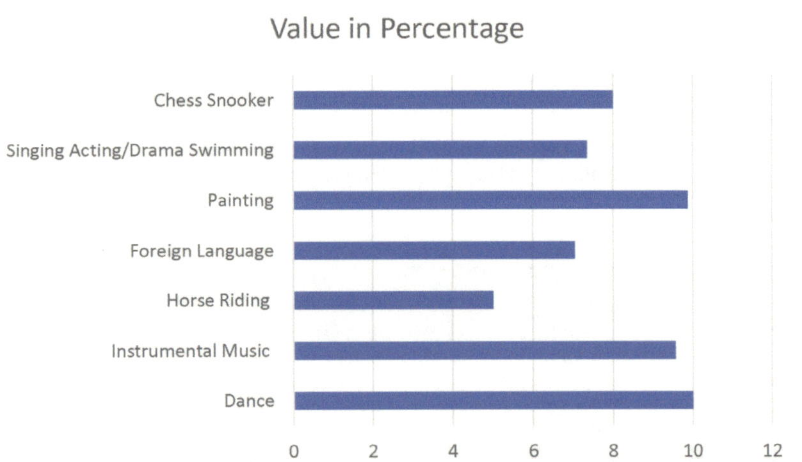

Figure 7.
Graphical representation.

	EQ	IQ	AQ	CQ
1 Year	16.34	13.34	12.21	21.94
2 Year	18.34	16.98	16.21	22.03
3 Year	21.21	19.2	19.21	22.94
4 Year	25.08	22.45	21.21	23.94
5 Year	26.78	25.08	24.21	23.94

Table 1.
Values of EQ, IQ, AQ, and CQ from fingerprint of children from 1 to 5 years.

4. Conclusion

Cognitive science, cognitive informatics, and computer modeling with pattern recognition of finger require some basic fundamentals for their implementation as cognitive concept in various applications of science and engineering, and the chapter has introduced and has bright future perspectives. It can be used as a useful adjunct to aid in preliminary study of field and behavior of child. Such measures can also help the couple to seek appropriate medical care and services for affected children. It will help the parents to be better equipped with management of such children. Also, the early detection of inborn errors is crucial because it can used as a vital tool to counsel the couple about avoiding conception of further affected fetuses. Historical background of cognitive science, cognitive map, and perception to conception were introduced in addition to cognitive network, modeling, and architecture for brain mapping with human fingerprint.

Author details

Rohit Raja[1]*, Hiral Raja[2], RajKumar Patra[3], Kamal Mehta[4], Akanksha Gupta[1] and Kunta Ramya Laxmi[5]

1 IT Department, GGV (A Central University), Bilaspur, Chhattisgarh, India

2 AIET Hyderabad, India

3 CSE Department, CMR College, Hyderabad, India

4 Computer Science and Information Technology, Mukesh Patel School of Technology Management and Engineering, NMIMS, Shirpur, India

5 CSE Department, SIET Hyderabad, India

*Address all correspondence to: drrohitraja1982@gmail.com

References

[1] Siddiqui AMN, Telgad R, Lothe S, Deshmukh PD. Fingerprint recognition system for person identification using termination and bifurcation minutiae. IOSR Journal of Computer Engineering (IOSR-JCE):33-38. 2015. e-ISSN: 2278-0661,p-ISSN: 2278-8727. Available from: www.iosrjournals.org

[2] Raja R, Sinha TS, Dubey RP. Recognition of human-face from side-view using progressive switching pattern and soft-computing technique. AMSE Journals—2015-Series: Advances B. 2015;**58**(1):14-34

[3] Ravi Subban and Dattatreya P. Mankame, A study of biometric approach using fingerprint recognition, Lecture Notes on Software Engineering. May 2013;**1**(2):209-213

[4] Hoang Thai and Ha Nhat Tam, Fingerprint recognition using standardized fingerprint model. IJCSI International Journal of Computer Science Issues, May 2010 , Vol. 7, Issue 3, No 7. ISSN (Online): 1694-0784 ISSN (Print): 1694-0814.

[5] Ali MMH, Mahale VH, Yannawar P, Gaikwad AT. Overview of fingerprint recognition system. In: International Conference on Electrical, Electronics, and Optimization Techniques (ICEEOT). Chennai. 2016. pp. 1334-1338. DOI: 10.1109/ICEEOT.2016.7754900

[6] Martin Sagayam K, Narain Ponraj D, Jenkin Winston YJC, Esther Jeba D, Clara A. Authentication of biometric system using fingerprint recognition with Euclidean distance and neural network classifier. International Journal of Innovative Technology and Exploring Engineering (IJITEE). 2019;**8**(4):766-771. ISSN: 2278-3075

[7] Nirmal SB, Kinage KS. Contactless fingerprint recognition and fingerprint spoof mitigation using CNN. International Journal of Recent Technology and Engineering (IJRTE). 2019;**8**(4): 9271-9275. ISSN: 2277-3878

[8] Raja R, Kumar S, Mahmood RM. Color object detection based image retrieval using ROI segmentation with multi-feature method. Wireless Personal Communications. 2020;**112**:169-192. DOI: 10.1007/s11277-019-07021-6

[9] Jadhav VV, Patil RR, Jadhav RC, Magikar AN. Efficient biometric authentication technique using fingerprint. International Journal of Computer Science and Information Technologies. 2016;**7**(3):1132-1135. ISSN: 0975-9646

[10] Rani P, Sharma IIP. A review paper on fingerprint identification system. International Journal of Advanced Research in Computer Science & Technology (IJARCST 2014). 2014;**2**(3):58-60. ISSN: 2347-8446 (Online) ISSN: 2347-9817 (Print)

[11] Raja R. Physiological trait-based biometrical authentication of human-face using LGXP and ANN techniques. Int. J. Information and Computer Security. 2018;**10**(2/3):303-320

[12] Yang W, Wang S, Hu J, Zheng G, Valli C. Security and accuracy of fingerprint-based biometrics: A review. Symmetry. 2019;**11**:141. DOI: 10.3390/sym11020141

[13] Jain AK, Prabhakar S, Pankanti S. On the similarity of identical twin fingerprints. Pattern Recognition. 2002;**35**(11):2653-2663

[14] Jain AK, Maltoni D, Maio D, Prabhakar S. Handbook of Fingerprint Recognition. New York: Springer; 2003. ISBN: 9781848822535

[15] S. Chikkerur, C. Wu and V. Govindaraju, "A systematic approach

for feature extraction in fingerprint images," in First International Conference, ICBA, Hong Kong, 2004.

[16] Bartunek JS, Nilsson M, Sallberg B, Claesson I. Adaptive fingerprint image enhancement with emphasis on preprocessing of data. IEEE Transactions on Image Processing. 2013;**22**(2):644-656

[17] Pieter P. Historical cognitive science—Analysis and examples [Dissertation of PG Diploma in Logic, History and Philosophy of Science]. Belgium: Ghent University; 2015

[18] Shih JJ, Krusienski DJ, Wolpaw JR. Brain-computer interfaces in medicine. Mayo Clinic Proceedings. 2012;**87**(3):268-279. DOI: 10.1016/j.mayocp.2011.12.008

[19] Basar E, Basar-Eroglu C, Karakas S, Schurmann M. Are cognitive processes manifested in event-related gamma, alpha, theta and delta oscillations in the EEG? Neuroscience Letters. 1999;**259**(3):165168

[20] Wiggins GA, Bhattacharya J. Mind the gap: An attempt to bridge computational and neuroscientific approaches to study creativity. Frontiers in Human Neuroscience. 2014;**8**:540555

[21] Hasson U, Nusbam HC. Emerging opportunities for advancing cognitive neuroscience. Trends in Cognitive Neuroscience. 2019;**1898**:13

[22] Grillner S, Ip N, Koch C, Koroshetz W, Okano H, Polachek M, et al. Worldwide initiatives to advance brain research. Nature Neuroscience. 2019;**19**:11181122

[23] Robert NM, Harvey W. Introduction: New frontiers in the cognitive science of religion. Journal of Cognition and Culture. 2005;**5**:113

[24] Stefanie M. Selective deployment of attention to time and modality and its impact upon behavior and brain oscillations [PhD thesis in Department of Experimental and Health Sciences]. Salvador: University of Barcelona; 2016. p. 242

[25] David GS. Models of memory: Wittgenstein and cognitive science. Philosophical Psychology. 1991;**4**(2):203218

[26] Silvia C. The multisensory visual cortex: Cross-modal shaping of visual cortical responses and perception [PhD thesis for Doctoral Program in Experimental Psychology, Linguistics and Cognitive Neuroscience]. Bicocca: University of Milano; 2014

[27] Iris VR. The tractable cognition thesis. Cognitive Science. 2008;**32**:939984

[28] Pulin A, Stan F, Javier S. Sensory memory for grounded representations in a cognitive architecture. ACS Poster Collect. 2018;**1**:118

[29] Włodzisław D. Neurocognitive informatics manifesto. Series of information and management sciences. In: 8th International Conference on Information and Management Sciences (IMS 2009). Kunming-Banna, Yunnan, China: California Polytechnic State University; 2009. p. 264282

[30] Adam CJ. On the use of cognitive maps [PhD dissertation]. Faculty of the Graduate School, University of Minnesota; 2008

[31] Elyana S. Interaction between visual perception and mental representations of imagery and memory in the early visual areas [PhD thesis]. Finland: Institute of Behavioral Sciences, University of Helsinki; 2005

[32] Salvador SF, Charles S, Donna L, Alan K. Moving multisensory research along: motion perception across sensory modalities. Current Directions in Psychological Science. 2001;**13**(1):2932